Domain-Driven Design with Java – A Practitioner's Guide

Create simple, elegant, and valuable software solutions for complex business problems

Premanand Chandrasekaran

Karthik Krishnan

BIRMINGHAM—MUMBAI

Domain-Driven Design with Java – A Practitioner's Guide

Associate Group Product Manager: Gebin George

Publishing Product Manager: Kushal Dave

Senior Editor: Rohit Singh

Content Development Editor: Kinnari Chohan

Technical Editor: Pradeep Sahu

Copy Editor: Safis Editing

Project Coordinator: Manisha Singh

Proofreader: Safis Editing

Indexer: Sejal Dsilva

Production Designer: Nilesh Mohite

Marketing Coordinator: Sonakshi Bubbar

First published: July 2022

Production reference: 2100822

Published by Packt Publishing Ltd.

Livery Place

35 Livery Street

Birmingham

B3 2PB, UK.

ISBN 978-1-80056-073-4

www.packt.com

To my mother, Bhanumathy, for all her sacrifices and for exemplifying the power of determination. To my wife, Vidya, for being my loving partner throughout our joint life journey and to our son, Siddharth, for being a constant source of joy and pride.

– Prem

To my late mother, Visalam, and my father, Krishnan, for inculcating the value of hard work and perseverance. To my loving wife, Suja, for being the patient partner and providing a lot of encouragement, and to my daughter, Ananya, for being the source of cheer and happiness in our family.

– Karthik

Forewords

The ideas behind **Domain-driven Design** (DDD) have an outsized influence on software architecture and the way we think about software projects. *Ubiquitous Language* is a term defined by DDD, describing the practice of creating a more precise language when discussing design in software projects. However, DDD itself has provided a ubiquitous language for software architecture: most discussions around microservices touch on terms originating in the DDD world.

However, the book that started it all, *Domain-driven Design* by Eric Evans, falls into the category of what Mark Twain called a classic: "a book that everyone wants to have read." The truth of the matter is that, while the book is packed full of influential ideas, it is quite abstract for readers looking for advice on how to build software. That doesn't indicate a deficiency in the source–it is designed to cover broad concepts, which encompasses many different facets of design, including but not limited to software development.

However, when developers need to utilize these ideas in a practical sense, such as in a concrete project, the abstract doesn't help–they need a practitioner's guide, which readers now hold in their hands. This book does an excellent job translating the concepts of DDD into useful overviews and concrete advice on how to implement the ideas. For example, a pertinent question that developers have that is unanswered in the DDD book: which is better to implement a project, imperative or functional programming style? Each has trade-offs; this book helps teams make those kinds of implementation decisions. This book covers multiple perspectives on applying DDD to Java projects, including foundational material about how DDD and agile engineering practices intersect in modern projects.

While many books exist that cover DDD and software architecture independently, this book does an excellent job mapping one to another, describing how the design influence of DDD maps to a number of different architecture topologies. For example, it describes how teams can reconcile bounded contexts with layered and other architecture styles.

Part one of this book covers many foundational details of the book, including technical and design details. *Part two* covers Real World DDD, covering how to apply the abstract ideas in a more concrete realm. It covers everything from how to discover and document the domain, to how to map real-world concepts into DDD concepts. For example, it does an excellent job describing aggregates, their design, and relationships.

Part three covers a topic dear to my heart as the author of *Building Evolutionary Architectures*, which is the evolution of software. Software products constantly evolve, including new features and capabilities. Therefore, software development isn't a static process–it must evolve along multiple dimensions to remain successful. This book contains a number of useful evolution patterns, covering both the domain and architecture characteristics.

Towards the ideal of creating a practitioner's guide, the book also covers engineering practicalities such as logging, testing, versioning, and a host of other practical considerations. Though these areas may fall outside the scope of the DDD book, teams need them.

DDD offers a host of conceptual ideas but lacks implementation details. This book serves as an excellent bridge between abstract and implementation. While the focus is on Java, the content is broad enough for any practitioner to get a good overview of the many important considerations for DDD projects. In fact, each specific technology stack would benefit from having a practitioner's guide such as this one.

- Neal Ford
Director/Software Architect / Meme Wrangler
Thoughtworks, Inc.

The early 2000s were a dynamic time in the software industry. A number of luminaries were exploring lightweight "agile" processes that had the effect of converting software engineering from a paint-by-numbers activity into first-class knowledge work. Agile methodologies such as eXtreme Programming were natural extensions of arguments put forth by Jack Reeves in his 1992 C++ Journal article *What is Software Design*, which claimed that source code – not design docs or architecture diagrams – was the only equivalent to design specifications found in other engineering disciplines. The split between design and manufacturing still exists, of course, but with software, we don't need people for manufacturing. The compiler does it for us, which means that all work the people do is design work, and agile processes pushed the new generation of software engineers into messy design conversations with "the business."

The turn of the century also corresponded with the dotcom boom (and subsequent bust) as user interactions shifted from desktop applications to the web, creating technical scaling problems that were orders of magnitude beyond what the industry had faced up to that point. Distributed technologies weren't exactly new, but the ever-increasing technical complexity needed to solve problems of scale required new architectural approaches and created a demand for software engineering talent.

It was in that environment that Eric Evans published his famous "blue book," introducing the concept of **Domain-driven Design** (**DDD**) in 2003. In it, Evans gave us a set of techniques that directly mapped the design aspects of software development onto the way we wrote source code and patterns that helped us manage technical complexity at scale.

It is precisely because DDD has aged so well that we forget how much the world has changed since it was first published. User behavior changes triggered by the introduction of smartphones pushed organizations to externalize integration interfaces that had previously been hidden inside an internal network, previously accessed perhaps by a public website hosted in a DMZ, but not by external users directly. Continuous delivery accelerated the pace of software change and, as a consequence, the pace of the design process. Microservices and modern event-driven architectures created architectural patterns that enabled more software design activities to happen in parallel as organizations looked to scale the throughput of delivery. Prem and Karthik have been on the leading edge of many of these changes, well-networked in the innovation crucible of Thoughtworks that helped name and evangelize some of these techniques, and directly applying them in large organizations eager to modernize.

DDD remains as relevant today as it did when Evans published his blue book because it directly confronts the root causes of why software design is hard: creating a shared understanding of the problem and modularizing the architecture in a way that communicates that understanding in code. Karthik and Prem share important updates on both fronts: techniques like Wardley mapping, EventStorming, and domain storytelling to create a shared understanding, and an updated view of how DDD patterns apply with modern architectural approaches. This is a book full of lessons learned in the trenches from experienced practitioners, with practical lessons for us all. I've personally sharpened my own understanding while reviewing the book, and I'm confident you will too.

- Brandon Byars
Head of Technology, Thoughtworks
Creator of Mountebank

Ever since my first encounter with **Domain-Driven Design (DDD)** in 2008, I have been convinced that this is how to approach the design of complex systems or systems in complex environments. While DDD gives an excellent foundation, I have found that the combination with **Command-Query Responsibility Separation (CQRS)** can be even more powerful, albeit challenging to implement.

The library of examples that I started to publish in 2009 evolved into Axon Framework, as early adopters began sharing their experiences using it. One of those early adopters was Prem, using it in a large and complex project at a financial institution.

With the rising popularity of microservices, DDD and CQRS have shown to be essential design principles. Someone once jokingly said: "there are three types of developers implementing microservices. Those who use DDD, those who don't realize they do, and those who fail." With the demand for more real-time systems and the rise in popularity of event-driven systems, Event Sourcing also gained more traction. This push for event-driven systems has led to several interesting design techniques that take these events as the very starting point for exploring a system's behavior.

While I consider the famous *Blue Book* by Eric Evans a masterpiece that aged very well, it also stays very much in the abstract. Over the years, concepts and practices have been refined and adapted to changes in how we use technology. These concepts and practices are described in numerous resources scattered across the internet, making them hard to find and distill for those venturing into these realms for the first time. Prem and Karthik do an outstanding job of taking you on the journey from the essential concepts of Domain-Driven Design, via the design practices, to the actual implementation of a system. Their complete and pragmatic approach makes this book an excellent starting point for anyone exploring new approaches to complex system design. It's the book I wish I had when I started my journey.

- Allard Buijze
CTO and Founder of AxonIQ

Creator of Axon Framework

Contributors

About the authors

Premanand Chandrasekaran is a technology leader and change agent, with a solid track record of leading large technology teams and helping businesses deliver mission-critical problems while exhibiting high internal and external quality. In the past two decades, he has had the pleasure of helping a variety of clients and domains, including financial services, online retailers, education, and healthcare startups. His specialties include technical innovation, architecture, continuous delivery, agile/iterative transformation, and employee development. When not fiddling with his trusty laptop, he spends time cutting vegetables, cooking, playing video games, and analyzing the nuances of the game of cricket.

"I would like to first and foremost thank my loving and patient wife, Vidya, and son, Siddharth, for their continued support, patience, and encouragement throughout the long process of writing this book. I would also like to thank my colleague and mentor, Gagan Madan, for constantly challenging me, pushing my limits, and inspiring me to achieve greater heights. Finally, my co-author, Karthik – without his perseverance and gentle prodding, it would have been very hard to finish this project. Lastly, my employer, Thoughtworks, for encouraging me to undertake this project and all my fellow Thoughtworkers for being a constant source of inspiration!"

Karthik Krishnan is a technology leader with over 25 years of experience in designing and building large-scale enterprise solutions across financial and retail domains. He has played numerous technical roles in leading product development for major financial institutions. He is currently serving the role of technical principal at Thoughtworks. He is passionate about platform thinking, solution architecture, and application security and strives to be known as a coding architect. His most recent assignment entailed leading a large technology team, helping their clients in their legacy modernization journey with the cloud. When not working, he spends time practicing playing tunes on his musical keyboard.

"I would like to thank my wife, Suja, and daughter, Ananya, for being my pillar of support, providing all the necessary encouragement, and, more importantly, for being very understanding, patient, and accommodating of my long book-writing sessions eating into their weekend plans. This book would not have been possible without them. And I would like to thank my friend, colleague, and co-author, Prem, for providing the energy and bringing in new ideas for discussion and collaborating with me, encouraging healthy debates and discussions throughout the creative process of this book. Lastly, my employer, Thoughtworks, for providing me with the space and encouraging me to write this book, and all my colleagues at Thoughtworks for providing their valuable feedback throughout the course of this book-writing journey."

About the reviewer

Viktor Daróczi is a software engineer and programming enthusiast. His journey as a developer started in the 1990s when he managed to get his hands on a used Commodore 64. From then on, he couldn't stop and been developing e-commerce platforms, crowd-testing solutions, an IoT platform, and interactive mobile ads, and ended up in fintech at PagoNxt (a Santander company).

Viktor lives in Munich, from where he initiates his explorations of the blue planet in the company of three beloved girls, his wife and two daughters, whom he tends to bore with talking about programming, math, and ancient human languages.

Table of Contents

Part 2: Real-World DDD

3

Understanding the Domain

4

Domain Analysis and Modeling

5

Implementing Domain Logic

6

Implementing the User Interface – Task-Based

7

Implementing Queries

8

Implementing Long-Running Workflows

9

Integrating with External Systems

Part 3: Evolution Patterns

10

Beginning the Decomposition Journey

11

Decomposing into Finer-Grained Components

12

Beyond Functional Requirements

Index

Other Books You May Enjoy

Preface

Domain-driven design (DDD) makes available a set of principles, patterns, and techniques that subject-matter experts, architects, developers, and other team members can adopt to work together and decompose complex systems into well-factored, collaborative, and loosely coupled subsystems. When Eric Evans introduced these concepts in the early 2000s, in a lot of ways, these principles were way ahead of their time. We were firmly in the age of the monolith, **service-oriented architectures** (**SOAs**) as a concept were just starting to take root, and the cloud, microservices, continuous delivery, and so on didn't even exist yet! While it was relatively easy to adopt some of its tactical aspects, the strategic side of DDD was still seen as an unjustifiable overhead for the most part.

Fast-forwarding to today, we are building our most complex software solutions ever, with even more complex organization and team structures to cope. Also, the use of the public cloud is almost a given. This has given rise to a situation where distributed teams and applications are almost a norm. Also, we are also in an age where applications from an earlier generation need to be modernized. All this has resulted in the principles of DDD, specifically the strategic elements, gaining a lot of prominence.

We have been practitioners of these concepts and have gained valuable insights from our experiences. Over the years, we have seen a number of advancements that have made the adoption of DDD at scale a viable option. This book is a distillation of all our collective experiences. While we have drawn a lot of inspiration from earlier works on the subject, we have been very conscious to apply a practitioner's mindset so that we lower the barrier for teams looking to sustain and thrive in their journey of building complex, distributed software.

Who this book is for

This book was written with a diverse set of roles and skills in mind. While the concepts of DDD have been in existence for a long time, practical application and scaling have been a challenge, arguably due to a dearth of practical techniques, tools, and real-world examples that bring all these concepts together as a cohesive whole. Successful application of these principles requires strong collaboration from a varied set of roles and disciplines across an organization, including executives, business experts, product owners, business analysts, architects, developers, testers, and operators.

Here is a quick summary of reader personas and what they will gain from reading this book:

Executives and business experts should read this book so that they can articulate their vision and the core concepts that justify the need for the solution. Techniques will allow them to do this in an expedient manner and also gain empathy toward what it takes to implement changes quickly and reliably.

Product owners should read this book so that they can act as effective facilitators when communicating with both business and technical team members, making sure that there is no loss in translation.

Architects should read this book so that they gain an appreciation of the fact that it is of utmost importance to understand the problem before thinking of a solution. They will also gain an appreciation of various architecture patterns and how they play in conjunction with DDD principles.

Developers and testers will be able to put their knowledge to work with this practical guide to create elegant software designs that are easy and pleasant to work with and reason about.

The book provides a hands-on approach to gathering requirements effectively, promoting a shared understanding among all team members in order to implement solutions that will be able to withstand the test of a dynamically evolving business ecosystem.

What this book covers

Chapter 1, The Rationale for Domain-Driven Design, examines how the practice of DDD provides a set of guidelines and techniques to improve the odds of success in our favor. We will look at how Eric Evans' classic book on the subject from 2003 is extremely relevant today. We will also introduce the elements of strategic and tactical DDD.

Chapter 2, Where and How Does DDD Fit?, examines how DDD compares with several of these architecture styles and how/where it fits in the overall scheme of things when crafting a software solution.

Chapter 3, Understanding the Domain, introduces the sample domain (International Trade) at a fictitious KP bank. We also examine how we can get started with strategic design using techniques like business model canvas, impact maps, and Wardley maps.

Chapter 4, Domain Analysis and Modeling, continues the analysis and modeling of the sample problem domain – Letter of Credit (LC) application, by using techniques like domain storytelling and eventstorming to arrive at a shared understanding of the problem and brainstorm ideas to arrive at a solution.

Chapter 5, Implementing Domain Logic, implements the command-side API for the sample application. We will look at how we can employ an event-driven architecture to build loosely coupled components. We will also look at how to implement structural and business validations and persistence options by contrasting state-stored and event-sourced aggregates.

Chapter 6, Implementing the User Interface – Task-Based, designs the **user interace** (**UI**) for the sample application. We will also express expectations of the UI to the service implementation.

Chapter 7, Implementing Queries, dives deeper into how we can construct read-optimized representations of data by listening to domain events. We will also look at persistence options for these read models.

Chapter 8, Implementing Long-Running Workflows, looks at implementing both long-running user operations (sagas) and deadlines. We will also look at how we can keep track of the overall flow using log aggregation and distributed tracing. We will round off by looking at when/whether to choose explicit orchestration components of implicit choreography.

Chapter 9, Integrating with External Systems, looks at integrating with other systems and bounded contexts. We will present the various styles of integration and the implications of choosing each of these.

Chapter 10, Beginning the Decomposition Journey, decomposes the command and the query side of the sample-bounded context into distinct components. We will look at the trade-offs involved when making these choices.

Chapter 11, Decomposing into Finer-Grained Components, looks at finer-grained decomposition and the trade-offs involved beyond the technical implications. We will decompose our application into distinct functions and discuss where it might be appropriate to draw the line.

Chapter 12, Beyond Functional Requirements, looks at factors beyond business requirements that can play a significant role in how applications are decomposed. Specifically, we will examine the effect that cross-functional requirements play when applying DDD.

To get the most out of this book

This book is aimed at a wide range of software team member personas. Some amount of prior experience building software solutions is assumed. Code examples in the book use the Java programming language. Familiarity with OO and frameworks such as Spring will be very helpful.

Software/hardware covered in the book	Operating system requirements
Spring Framework	Windows, macOS, or Linux
Axon Framework	
JavaFX	

If you are using the digital version of this book, we advise you to type the code yourself or access the code from the book's GitHub repository (a link is available in the next section). Doing so will help you avoid any potential errors related to the copying and pasting of code.

Download the example code files

You can download the example code files for this book from GitHub at `https://github.com/PacktPublishing/Domain-Driven-Design-with-Java-A-Practitioner-s-Guide`. If there's an update to the code, it will be updated in the GitHub repository.

We also have other code bundles from our rich catalog of books and videos available at `https://github.com/PacktPublishing/`. Check them out!

Download the color images

We also provide a PDF file that has color images of the screenshots and diagrams used in this book. You can download it here:

`https://packt.link/TwzEB.`

Conventions used

There are a number of text conventions used throughout this book.

`Code in text`: Indicates code words in text, database table names, folder names, filenames, file extensions, pathnames, dummy URLs, user input, and Twitter handles. Here is an example: "As shown in the event storming artifact, the LC Application aggregate is able to handle ApproveLCApplicationCommand, which results in LCApplicationApprovedEvent."

A block of code is set as follows:

```
1   import org.axonframework.deadline.DeadlineManager;
2
3   class LCApplication {
4       //...
5       @CommandHandler
6       public void on(SubmitLCApplicationCommand command,
7                       DeadlineManager deadlineManager) { ①
8           assertPositive(amount);
9           assertMerchandise(merchandise);
10          assertInDraft(state);
11          apply(new LCApplicationSubmittedEvent(id, amount));
12
13          deadlineManager.schedule(Duration.ofDays(10), ②
14              "LC_APPROVAL_REMINDER",
15              LCApprovalPendingNotification.first(id)); ③
16      }
17      //...
18  }
```

> **Tips or Important Notes**
> Appear like this.

Get in touch

Feedback from our readers is always welcome.

General feedback: If you have questions about any aspect of this book, email us at `customercare@ packtpub.com` and mention the book title in the subject of your message.

Errata: Although we have taken every care to ensure the accuracy of our content, mistakes do happen. If you have found a mistake in this book, we would be grateful if you would report this to us. Please visit www.packtpub.com/support/errata and fill in the form.

Piracy: If you come across any illegal copies of our works in any form on the internet, we would be grateful if you would provide us with the location address or website name. Please contact us at copyright@packt.com with a link to the material.

If you are interested in becoming an author: If there is a topic that you have expertise in and you are interested in either writing or contributing to a book, please visit authors.packtpub.com.

Share Your Thoughts

Once you've read *Domain-Driven Design with Java - A Practitioner's Guide*, we'd love to hear your thoughts! Scan the QR code below to go straight to the Amazon review page for this book and share your feedback.

https://packt.link/r/1800560737

Your review is important to us and the tech community and will help us make sure we're delivering excellent quality content.

Part 1: Foundations

While the IT industry prides itself on being at the very bleeding edge of technology, it also oversees a relatively high proportion of projects that fail outright or do not meet their originally intended goals for one reason or another. In *Part 1*, we will look at the reasons for software projects not achieving their intended objectives and how practicing **Domain-Driven Design (DDD)** can significantly help improve the odds of achieving success. We will do a quick tour of the main concepts that Eric Evans elaborated on in his seminal book of the same name and examine why/how it is extremely relevant in the age of distributed systems. We will also look at several popular architecture styles and programming paradigms and explore how DDD fits in the scheme of things.

This part contains the following chapters:

- *Chapter 1, The Rationale for Domain-Driven Design*
- *Chapter 2, Where and How Does DDD Fit?*

The Rationale for Domain-Driven Design

The being cannot be termed rational or virtuous, who obeys any authority, but that of reason.

— *Mary Wollstonecraft*

According to the **Project Management Institute's (PMI's)** *Pulse of the Profession* report published in February 2020, only 77% of all projects meet their intended goals—and even this is true only in the most mature organizations. For less mature organizations, this number falls to just 56%; that is, approximately one in every two projects does not meet its intended goals. Furthermore, approximately one in every five projects is declared an outright failure. At the same time, we also seem to be embarking on our most ambitious and complex projects.

In this chapter, we will examine the main causes of project failure and look at how applying **domain-driven design** (DDD) provides a set of guidelines and techniques to improve the odds of success in our favor. While Eric Evans wrote his classic book on the subject way back in 2003, we look at why that work is still extremely relevant in today's times.

In this chapter, we will cover the following topics:

- Understanding why software projects fail

- Characteristics of modern systems and dealing with complexity

- Introduction to DDD

- Reviewing why DDD is relevant today

By the end of this chapter, you will have gained a basic understanding of DDD and why you should strongly consider applying the tenets of DDD when architecting/implementing modern software applications, especially the more complex ones.

Why do software projects fail?

Failure is simply the opportunity to begin again, this time more intelligently.

— Henry Ford

According to the project success report published in the *Project Management Journal* of the PMI, the following six factors need to be true for a project to be deemed successful:

Category	Criterion	Description
Project	Time	It meets the desired time schedules.
	Cost	Its cost does not exceed the budget.
	Performance	It works as intended.
Client	Use	Its intended clients use it.
	Satisfaction	Its intended clients are happy.
	Effectiveness	Its intended clients derive direct benefits through its implementation.

Table 1.1 – Project success factors

With all of these criteria being applied to assess project success, a large percentage of projects fail for one reason or another. Let's examine some of the top reasons in more detail.

Inaccurate requirements

PMI's *Pulse of the Profession* report from 2017 highlights a very stark fact—a vast majority of projects fail due to inaccurate or misinterpreted requirements. It follows that it is impossible to build something that clients can use, they are happy with, and that makes them more effective at their jobs if the wrong thing gets built—even much less for the project to be built on time and within budget.

IT teams, especially in large organizations, are staffed with mono-skilled roles, such as UX designer, developer, tester, architect, business analyst, project manager, product owner, and business sponsor. In a lot of cases, these people are parts of distinct organization units/departments—each with its own set of priorities and motivations. To make matters even worse, the geographical separation between these people only keeps increasing. The need to keep costs down and the recent COVID-19 ecosystem does not help matters either.

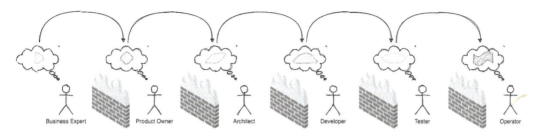

Figure 1.1 – Silo mentality and the loss of information fidelity

All this results in a loss in fidelity of information at every stage in the assembly line, which then results in misconceptions, inaccuracies, delays, and eventually failure!

Too much architecture

Writing complex software is quite a task. You cannot just hope to sit down and start typing code—although that approach might work in some trivial cases. Before translating business ideas into working software, a thorough understanding of the problem at hand is necessary. For example, it is not possible (or is at least extremely hard) to build credit card software without understanding how credit cards work in the first place. To communicate your understanding of a problem, it is not uncommon to create software models of the problem before writing code. This model or collection of models represents the understanding of the problem and the architecture of the solution.

Efforts to create a perfect model of the problem—one that is accurate in a very broad context—are not dissimilar to the proverbial holy grail quest. Those accountable for producing the architecture can get stuck in analysis paralysis and/or big design upfront, producing artifacts that are one or more of too high level, wishful, gold-plated, buzzword-driven, or disconnected from the real world—while not solving any real business problems. This kind of lock-in can be especially detrimental during the early phases of the project when the knowledge levels of team members are still up and coming. Needless to say, projects adopting such approaches find it hard to reach success consistently.

> **Tip**
> For a more comprehensive list of modeling anti-patterns, refer to Scott W. Ambler's website (`http://agilemodeling.com/essays/enterpriseModelingAntiPatterns.htm`) and book, *Agile Modeling: Effective Practices for eXtreme Programming and the Unified Process*, dedicated to the subject.

Too little architecture

Agile software delivery methods manifested themselves in the late 90s and early 2000s in response to heavyweight processes collectively known as *waterfall*. These processes seemed to favor big design upfront and abstract ivory tower thinking based on wishful, ideal-world scenarios. This was based on the premise that thinking things out well in advance ends up saving serious development headaches later on as the project progresses.

In contrast, agile methods seem to favor a much more nimble and iterative approach to software development with a high focus on working software over other artifacts, such as documentation. Most teams these days claim to practice some form of iterative software development. However, with this obsession to claim conformance to a specific family of agile methodologies as opposed to the underlying principles, a lot of teams misconstrue having just enough architecture with having no perceptible architecture. This results in a situation where adding new features or enhancing existing ones takes a lot longer than what it previously used to—which then accelerates the devolution of the solution to become the dreaded big ball of mud (`http://www.laputan.org/mud/mud.html#BigBallOfMud`).

Excessive incidental complexity

Mike Cohn popularized the notion of the test pyramid, where he talks about how a large number of unit tests should form the foundation of a sound testing strategy—with numbers decreasing significantly as you move up the pyramid. The rationale here is that as you move up the pyramid, the cost of upkeep goes up copiously while the speed of execution slows down manifold. In reality, though, a lot of teams seem to adopt a strategy that is the exact opposite of this—known as the testing ice cream cone, as depicted here:

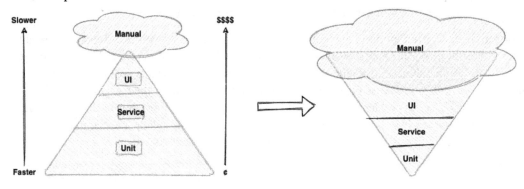

Figure 1.2 – Testing strategy: expectation versus reality

The testing ice cream cone is a classic case of what Fred Brooks calls incidental complexity in his seminal paper titled *No Silver Bullet—Essence and Accident in Software Engineering* (`http://worrydream.com/refs/Brooks-NoSilverBullet.pdf`). All software has some amount of essential complexity that is inherent to the problem being solved. This is especially true when creating solutions for non-trivial problems. However, incidental or accidental complexity is not directly attributable to the problem itself—but is caused by the limitations of the people involved, their skill levels, the tools, and/or abstractions being used. Not keeping tabs on incidental complexity causes teams to veer away from focusing on the real problems, solving which provide the most value. It naturally follows that such teams minimize their odds of success appreciably.

Uncontrolled technical debt

Financial debt is the act of borrowing money from an outside party to quickly finance the operations of a business—with the promise to repay the principal plus the agreed-upon rate of interest in a timely manner. Under the right circumstances, this can accelerate the growth of a business considerably while allowing the owner to retain ownership, reduced taxes, and lower interest rates. On the other hand, the inability to pay back this debt on time can adversely affect credit rating, result in higher interest rates, cash flow difficulties, and other restrictions.

Technical debt is what results when development teams take arguably suboptimal actions to expedite the delivery of a set of features or projects. For a period of time, just like borrowed money allows you to do things sooner than you could otherwise, technical debt can result in short-term speed. In the long term, however, software teams will have to dedicate a lot more time and effort toward simply managing complexity as opposed to thinking about producing architecturally sound solutions. This can result in a vicious negative cycle, as illustrated in the following diagram:

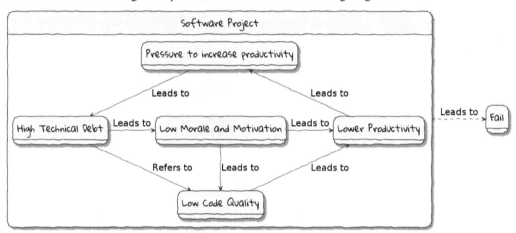

Figure 1.3 – Technical debt: implications

In a recent McKinsey survey (https://www.mckinsey.com/business-functions/ mckinsey-digital/our-insights/tech-debt-reclaiming-tech-equity) sent out to CIOs, around 60% reported that the amount of technical debt increased over the past 3 years. At the same time, over 90% of CIOs allocated less than a fifth of their tech budget toward paying it off. Martin Fowler explores (https://martinfowler.com/articles/is-quality- worth-cost.html#WeAreUsedToATrade-offBetweenQualityAndCost) the deep correlation between high software quality (or the lack thereof) and the ability to enhance software predictably. While carrying a certain amount of technical debt is inevitable and part of doing business, not having a plan to systematically pay off this debt can have significantly detrimental effects on team productivity and the ability to deliver value.

Ignoring non-functional requirements

Stakeholders often want software teams to spend the majority (if not all) of their time working on features that provide enhanced functionality. This is understandable given that such features provide the highest ROI. These features are called functional requirements.

Non-functional requirements (also sometimes known as cross-functional requirements), on the other hand, are those aspects of the system that do not affect functionality directly but have a profound effect on the efficacy of those using and maintaining these systems. There are many kinds of NFRs. A partial list of common NFRs is depicted in the following figure:

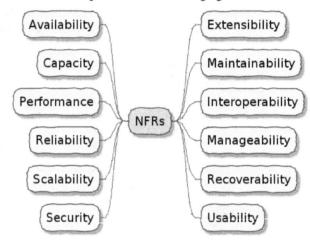

Figure 1.4 – NFRs

Very rarely do users explicitly request NFRs, but they almost always expect these features to be part of any system they use. Oftentimes, systems may continue to function without NFRs being met, but not without having an adverse impact on the *quality* of the user experience. For example, the home page of a website that loads in under 1 second under low load and takes upward of 30 seconds under higher loads may not be usable during those times of stress. Needless to say, not treating NFRs with the same amount of rigor as explicit, value-adding functional features can lead to unusable systems—and subsequently failure.

In this section, we examined some common reasons why software projects to fail. Is it possible to improve our odds? Before we do that, let's look at the nature of modern software systems and how we can deal with the ensuing complexity.

Modern systems and dealing with complexity

We cannot solve our problems with the same level of thinking that created them.

— Albert Einstein

As we saw in the previous section, there are several reasons why software endeavors fail. In this section, we will try to understand how software gets built, what the currently prevailing realities are, and what adjustments we need to make in order to cope.

How software gets built

Building successful software is an iterative process of constantly refining knowledge and expressing it in the form of models. We have attempted to capture the essence of the process at a high level here:

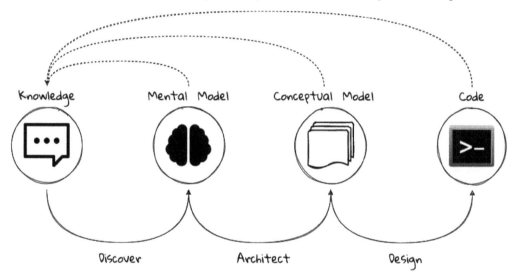

Figure 1.5 – Building software is a continuous refinement of knowledge and models

Before we express a solution in working code, it is necessary to understand *what* the problem entails, *why* the problem is important to solve, and finally, *how* it can be solved. Irrespective of the methodology used (waterfall, agile, and/or anything in between), the process of building software is one where we need to constantly use our knowledge to refine mental/conceptual models to be able to create valuable solutions.

Complexity is inevitable

We find ourselves in the midst of the fourth industrial revolution, where the world is becoming more and more digital—with technology being a significant driver of value for businesses. There have been exponential advances in computing technology, as illustrated by Moore's law:

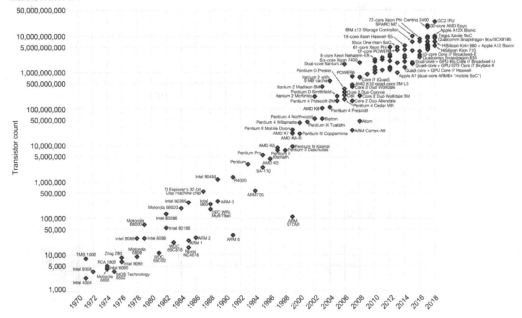

Moore's Law – The number of transistors on integrated circuit chips (1971-2018)

Moore's law describes the empirical regularity that the number of transistors on integrated circuits doubles approximately every two years. This advancement is important as other aspects of technological progress – such as processing speed or the price of electronic products – are linked to Moore's law.

Data source: Wikipedia (https://en.wikipedia.org/wiki/Transistor_count)
The data visualization is available at OurWorldinData.org. There you find more visualizations and research on this topic.

Licensed under CC-BY-SA by the author Max Roser.

Figure 1.6 – Moore's law

This has also coincided with the rise of the internet.

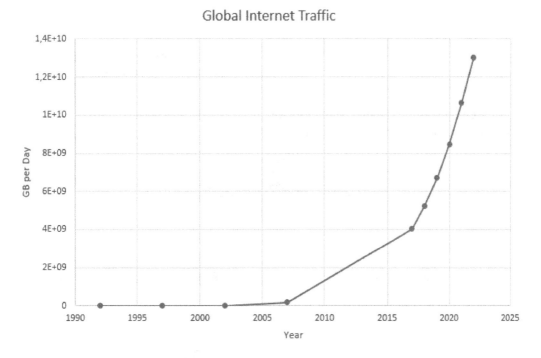

Figure 1.7 – Global internet traffic

This has meant that companies are being required to modernize their software systems much more rapidly than they ever have. Along with all this, the onset of commodity computing services, such as the public cloud, has led to a move away from expensive centralized computing systems to more distributed computing ecosystems. As we attempt to build our most complex solutions, monoliths are being replaced by an environment of distributed, collaborating microservices. Modern philosophies and practices, such as automated testing, architecture fitness functions, continuous integration, continuous delivery, DevOps, security automation, and infrastructure as code, to name a few, are disrupting the way we deliver software solutions.

All these advances introduce their own share of complexity. Instead of attempting to control the amount of complexity, there is a need to embrace and cope with it.

Optimizing the feedback loop

As we enter an age of encountering our most complex business problems, we need to embrace new ways of thinking, development philosophy, and an arsenal of techniques to iteratively evolve mature software solutions that will stand the test of time. We need better ways of communicating, analyzing problems, arriving at a collective understanding, creating and modeling abstractions, and then implementing and enhancing the solution.

To state the obvious—we're all building software with seemingly brilliant business ideas on one side and our ever-demanding customers on the other, as shown here:

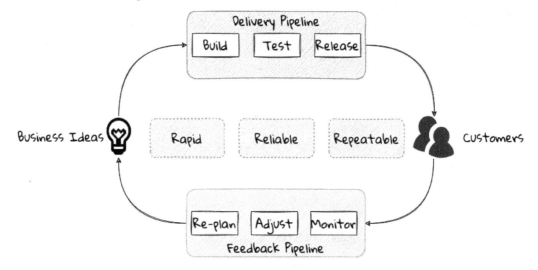

Figure 1.8 – The software delivery continuum

In between, we have two chasms to cross—the *delivery pipeline* and the *feedback pipeline*. The delivery pipeline enables us to put software in the hands of our customers, whereas the feedback pipeline allows us to adjust and adapt. As we can see, this is a continuum. If we are to build better, more valuable software, this continuum, this potentially infinite loop, has to be optimized!

To optimize this loop, we need three characteristics to be present: we need to be fast, we need to be reliable, and we need to do this over and over again. In other words, we need to be rapid, reliable, and repeatable—all at the same time! Take any one of these away and it just won't sustain.

DDD promises to provide answers on how to do this in a systematic manner. In the upcoming section, and indeed the rest of this book, we will examine what DDD is and why it is indispensable when working to provide solutions for non-trivial problems in today's world of massively distributed teams and applications.

What is DDD?

Life is really simple, but we insist on making it complicated.

— Confucius

In the previous section, we saw how a myriad of reasons, coupled with system complexity, gets in the way of software project success. The idea of DDD, originally conceived by Eric Evans in his 2003 book, is an approach to software development that focuses on expressing software solutions in the form of a model that closely embodies the core of the problem being solved. It provides a set of principles and systematic techniques to analyze, architect, and implement software solutions in a manner that enhances the chances of success.

While Evans' work is indeed seminal, ground-breaking, and way ahead of its time, it is not prescriptive at all. This is a strength in that it has enabled the evolution of DDD beyond what Evans had originally conceived at the time. On the other hand, it also makes it extremely hard to define what DDD actually encompasses, making practical application a challenge. In this section, we will look at some foundational terms and concepts behind DDD. Elaboration and practical application of these concepts will happen in upcoming chapters of this book.

When encountering a complex business problem, DDD suggests doing the following:

- **Understand the problem**: To have a deep, shared understanding of the problem, it is necessary for business and technology experts to collaborate closely. Here, we collectively understand what the problem is and why it is valuable to solve. This is termed the **domain** of the problem.

- **Break down the problem into more manageable parts**: To keep complexity at manageable levels, break down complex problems into smaller, independently solvable parts. These parts are termed **subdomains**. It may be necessary to further break down subdomains where the subdomain is still too complex. Assign explicit boundaries to limit the functionality of each subdomain. This boundary is termed the **bounded context** for that subdomain. It may also be convenient to think of the subdomain as a concept that makes more sense to the domain experts (in the problem space), whereas the bounded context is a concept that makes more sense to the technology experts (in the solution space).

- For each of these bounded contexts, do the following:

  - **Agree on a shared language**: Formalize the understanding by establishing a shared language that is applicable unambiguously within the bounds of the subdomain. This shared language is termed the ubiquitous language of the domain.

  - **Express understanding in shared models**: In order to produce working software, express the ubiquitous language in the form of shared models. This model is termed the **domain model**. There may exist multiple variations of this model, each meant to clarify a specific aspect of the solution, for example, a process model, a sequence diagram, working code, and a deployment topology.

- **Embrace the incidental complexity of the problem**: It is important to note that it is not possible to shy away from the essential complexity of a given problem. By breaking down the problem into subdomains and bounded contexts, we are attempting to distribute it (more or less) evenly across more manageable parts.

- **Continuously evolve for greater insight**: It is important to understand that the previous steps are not a one-time activity. Businesses, technologies, processes, and our understanding of these evolve, so it is important for our shared understanding to remain in sync with these models through continuous refactoring.

A pictorial representation of the essence of DDD is expressed here:

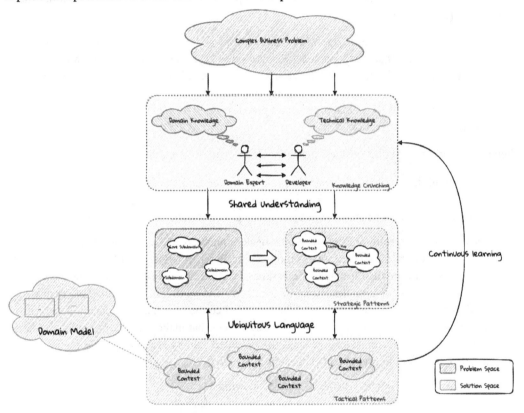

Figure 1.9 – Essence of DDD

We appreciate that this is quite a whirlwind introduction to the subject of DDD.

Understanding the problem using strategic design

In this section, let's demystify some commonly used concepts and terms when working with DDD. First and foremost, we need to understand what we mean by the first D—**domain**.

What is a domain?

The foundational concept when working with DDD is the notion of a domain. But what exactly is a domain? The word domain, which has its *origins* in the 1600s from the old French word *domaine* (power) and the Latin word *dominium* (property, right of ownership) is a rather confusing word. Depending on who, when, where, and how it is used, it can mean different things.

Noun [edit]

domain (*plural* **domains**)

1. A geographic **area** owned or controlled by a single person or organization. [quotations ▼]
   *The king ruled his **domain** harshly.*
2. A field or sphere of activity, influence or expertise.
   *Dealing with complaints isn't really my **domain**: get in touch with customer services.*
   *His **domain** is English history.*
3. A group of related items, topics, or subjects. [quotations ▼]
4. (*mathematics*) The set of all possible mathematical entities (points) where a given function is defined.
5. (*mathematics, set theory*) The set of input (argument) values for which a function is defined.
6. (*mathematics*) A ring with no zero divisors; that is, in which no product of nonzero elements is zero.
   Hyponym: integral domain
7. (*mathematics, topology, mathematical analysis*) An open and connected set in some topology. For example, the interval (0,1) as a subset of the real numbers.
8. (*computing, Internet*) Any DNS domain name, particularly one which has been delegated and has become representative of the delegated domain name and its subdomains. [quotations ▼]
9. (*computing, Internet*) A collection of DNS or DNS-like domain names consisting of a delegated domain name and all its subdomains.
10. (*computing*) A collection of information having to do with a domain, the computers named in the domain, and the network on which the computers named in the domain reside.
11. (*computing*) The collection of computers identified by a domain's domain names.
12. (*physics*) A small region of a magnetic material with a consistent magnetization direction.
13. (*computing*) Such a region used as a data storage element in a bubble memory.
14. (*data processing*) A form of technical metadata that represent the type of a data item, its characteristics, name, and usage. [quotations ▼]
15. (*taxonomy*) The highest rank in the classification of organisms, above kingdom; in the three-domain system, one of the taxa *Bacteria*, *Archaea*, or *Eukaryota*.
16. (*biochemistry*) A folded section of a protein molecule that has a discrete function; the equivalent section of a chromosome

Figure 1.10 – The meaning of domain changes with context

In the context of a business, however, the word domain covers the overall scope of its primary activity— the service it provides to its customers. This is also referred to as the **problem domain**. For example, Tesla operates in the domain of electric vehicles, Netflix provides online movies and shows, and McDonald's provides fast food. Some companies such as Amazon provide services in more than one domain—online retail and cloud computing, among others. The domain of a business (at least the successful ones) almost always encompasses fairly complex and abstract concepts. To cope with this complexity, it is usual to decompose these domains into more manageable pieces called subdomains. Let's understand subdomains in more detail next.

What is a subdomain?

In its essence, DDD provides means to tackle complexity. Engineers do this by breaking down complex problems into more manageable ones called **subdomains**. This facilitates better understanding and makes it easier to arrive at a solution. For example, the online retail domain may be divided into subdomains such as product, inventory, rewards, shopping cart, order management, payments, and shipping, as shown in the following figure:

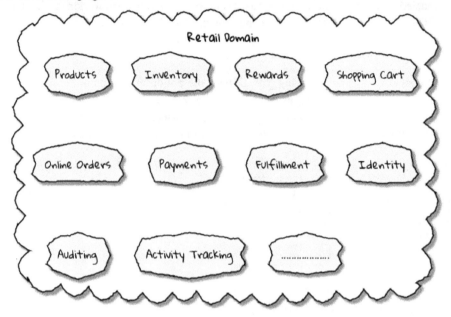

Figure 1.11 – Subdomains in the retail

In certain businesses, subdomains themselves may turn out to become very complex on their own and may require further decomposition. For instance, in the retail example, it may be required to break the products subdomain into further constituent subdomains, such as catalog, search, recommendations, and reviews, as shown here:

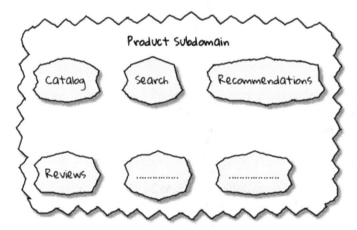

Figure 1.12 – Subdomains in the products subdomain

Further breakdown of subdomains may be needed until we reach a level of manageable complexity. Domain decomposition is an important aspect of DDD. Let's look at the types of subdomains to understand this better.

> **Important Note**
>
> The terms domain and subdomain tend to get used interchangeably quite often. This can be confusing to the casual onlooker. Given that subdomains tend to be quite complex and hierarchical, a subdomain can be a domain in its own right.

Types of subdomains

Breaking down a complex domain into more manageable subdomains is a great thing to do. However, not all subdomains are created equal. With any business, the following three types of subdomains are going to be encountered:

- **Core**: The main focus area for the business. This is what provides the biggest differentiation and value. It is, therefore, natural to want to place the most focus on the core subdomain. In the retail example, shopping carts and orders might be the biggest differentiation—and hence may form the core subdomains for that business venture. It is prudent to implement core subdomains in-house, given that it is something that businesses will desire to have the most control over. In the online retail example, the business may want to focus on providing an enriched experience for placing online orders. This will make the *online orders* and *shopping cart* part of the core subdomain.

- **Supporting**: Like with every great movie, where it is not possible to create a masterpiece without a solid supporting cast, so it is with supporting or auxiliary subdomains. Supporting subdomains are usually very important and very much required but may not be the primary focus of running the business. These supporting subdomains, while necessary to run the business, do not usually offer a significant competitive advantage. Hence, it might even be fine to completely outsource this work or use an off-the-shelf solution as is or with minor tweaks. For the retail example, assuming that online ordering is the primary focus of this business, catalog management may be a supporting subdomain.

- **Generic**: When working with business applications, you are required to provide a set of capabilities *not* directly related to the problem being solved. Consequently, it might suffice to just make use of an off-the-shelf solution. For the retail example, the identity, auditing, and activity tracking subdomains might fall in that category.

> **Important Note**
>
> It is important to note that the notion of core versus supporting versus generic subdomains is very context-specific. What is core for one business may be supporting or generic for another. Identifying and distilling the core domain requires a deep understanding and experience of what problem is being attempted to be solved.

Given that the core subdomain establishes most of the business differentiation, it will be prudent to devote the most amount of energy toward maintaining that differentiation. This is illustrated in the core domain chart here:

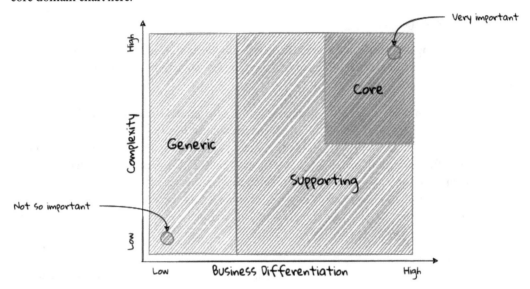

Figure 1.13 – Importance of subdomains

Over time, it is only natural that competitors will attempt to emulate your successes. Newer, more efficient methods will arise, reducing the complexity involved and disrupting your core. This may cause the notion of what is currently core to shift and become a supporting or generic capability, as depicted here:

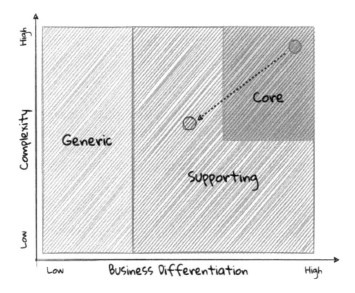

Figure 1.14 – Core domain erosion

To continue running a successful operation, it is required to constantly innovate at the core. For example, when AWS started the cloud computing business, it only provided simple infrastructure (IaaS) solutions. However, as competitors such as Microsoft and Google started to catch up, AWS has had to provide several additional value-added services (for example, PaaS and SaaS).

As is evident, this is not just an engineering problem. It requires a deep understanding of the underlying business. That's where domain experts can play a significant role.

Domain and technical experts

Any modern software team requires expertise in at least two areas—the functionality of the domain and the art of translating it into high-quality software. In most organizations, these exist as at least two distinct groups of people:

- **Domain experts**: Those who have a deep and intimate understanding of the domain. Domain experts are **subject-matter experts** (**SMEs**) who have a very strong grasp of the business. Domain experts may have varying degrees of expertise. Some SMEs may choose to specialize in specific subdomains, while others may have a broader understanding of how the overall business works.

- **Technical experts**: On the other hand, enjoy solving specific, quantifiable computer science problems. Often, technical experts do not feel it is worth their while to understand the context of the business they work in. Rather, they seem overly eager to only enhance their technical skills that are a continuation of their learnings in academia.

While the domain experts specify the *why* and the *what*, technical experts (software engineers) largely help realize the *how*. Strong collaboration and synergy between both groups are essential to ensure sustained high performance and success.

A divide originating in language

While strong collaboration between these groups is necessary, it is important to appreciate that these groups of people seem to have distinct motivations and differences in thinking. Seemingly, this may appear to be restricted to simple things such as differences in their day-to-day language. However, deeper analysis usually reveals a much larger divide in aspects such as goals and motivations. This is illustrated in the figure here:

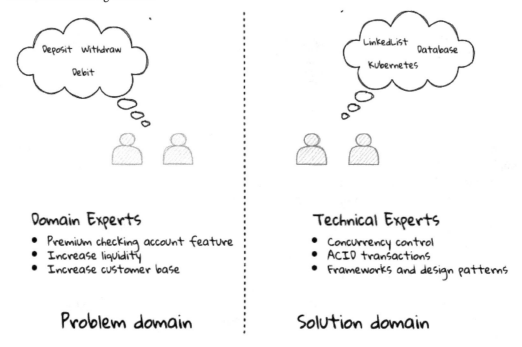

Figure 1.15 – Divide originating in language

But this is a book primarily focused on technical experts. Our point is that it is not possible to be successful by just working on technically challenging problems without gaining a sound understanding of the underlying business context.

Every decision we take regarding the organization, be it requirements, architecture, or code, has business and user consequences. In order to conceive, architect, design, build and evolve software effectively, our decisions need to aid in creating the optimal business impact. As mentioned, this can only be achieved if we have a clear understanding of the problem we intend to solve. This leads us to the realization that there exist two distinct *domains* when arriving at the solution for a problem.

> **Note**
>
> The use of the word domain in this context is done so in an abstract sense—not to be confused with the concept of the business domain introduced earlier.

Problem domain

This is a term that is used to capture information that simply defines the problem while consciously avoiding any details of the solution. It includes details such as *why* we are trying to solve the problem, *what* we are trying to achieve, and *how* it needs to be solved. It is important to note that the *why*, *what*, and *how* are from the perspective of the customers/stakeholders, not from the perspective of the engineers providing software solutions to the problem.

Consider the example of a retail bank that already provides a checking account capability for their customers. They want access to more liquid funds. They need to encourage customers to maintain higher account balances to achieve that. They are looking to introduce a new product called the *premium checking account* with additional features such as higher interest rates, overdraft protection, and no-charge ATM access. The problem domain expressed in the form of why, what, and how is shown here:

Question	Answer
Why	Bank needs access to more liquid funds
What	Have customers maintain higher account balances
How	By introducing a new product—the premium checking account with enhanced features

Table 1.2 – Problem domain: why, what, and how

Now that we have defined the problem and the motivations surrounding it, let's examine how it can inform the solution.

Solution domain

A term used to describe the environment in which the solution is developed. In other words, the process of translating requirements into working software (this includes design, development, testing, and deployment). Here, the emphasis is on the *how* of the problem being solved from a software implementation perspective. However, it is very difficult to arrive at a solution without having an appreciation of the why and the what.

Building on the previous premium checking account example, the code-level solution for this problem may look something like this:

```
 1 | class PremiumCheckingAccountFactory {
 2 |
 3 |     Account openPremiumCheckingAccount(Applicant applicant,
 4 |                                         MonetaryAmount initialAmount) {
 5 |
 6 |         Salary salary = salaryFor(applicant);
 7 |
 8 |         if (salary.isBelowThreshold()) {
 9 |             throw new InsufficientIncomeException(applicant);
10 |         }
11 |
12 |         Account account = Account.createFor(applicant);
13 |         account.deposit(initialAmount);
14 |         account.activate();
15 |         return account;
16 |     }
17 | }
```

This likely appears like a significant leap from a problem domain description, and indeed it is. Before a solution like this can be arrived at, there may need to exist multiple levels of refinement of the problem. This process of refinement is usually messy and may lead to inaccuracies in the understanding of the problem, resulting in a solution that may be good (for example, one that is sound from an engineering, software architecture standpoint) but not one that solves the problem at hand. Let's look at how we can continuously refine our understanding by closing the problem and solution domain gap.

Promoting a shared understanding using a ubiquitous language

Previously, we saw how organizational silos could result in valuable information getting diluted. At a credit card company I used to work with, the words plastic, payment instrument, account, **PAN** (**Primary Account Number**), **BIN** (**Bank Identification Number**), and card were all used by different team members to mean the exact same thing—the *credit card*—when working in the same area of the application. On the other hand, a term such as *user* would be used to sometimes mean a customer, a relationship manager, or a technical customer support employee. To make matters worse, a lot of these muddled use of terms got implemented in code as well. While this might feel like a trivial thing, it had far-reaching consequences. Product experts, architects, and developers all came and went, each regressively contributing to more confusion, muddled designs, implementation, and technical debt with every new enhancement—accelerating the journey toward the dreaded, unmaintainable big ball of mud (http://www.laputan.org/mud/).

DDD advocates breaking down these artificial barriers and putting the domain experts and the developers on the same level footing by working collaboratively toward creating what DDD calls a *ubiquitous language*—a shared vocabulary of terms, words, and phrases to continuously enhance the collective understanding of the entire team. This phraseology is then used actively in every aspect of the solution: the everyday vocabulary, the designs, the code—in short, by *everyone* and *everywhere*. Consistent use of the common, ubiquitous language helps reinforce a shared understanding and produce solutions that better reflect the mental model of the domain experts.

Evolving a domain model and a solution

The ubiquitous language helps establish a consistent, albeit informal, lingo among team members. To enhance understanding, this can be further refined into a formal set of abstractions—a *domain model* to represent the solution in software. When a problem is presented to us, we subconsciously attempt to form mental representations of potential solutions. Furthermore, the type and nature of these representations (models) may differ wildly based on factors such as our understanding of the problem, our backgrounds, and experiences. This implies that it is natural for these models to be different. For example, the same problem can be thought of differently by various team members, as shown here:

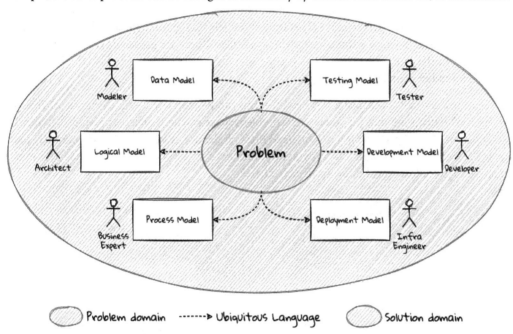

Figure 1.16 – Multiple models to represent the solution

As illustrated here, the business expert may think of a process model, whereas the test engineer may think of exceptions and boundary conditions to arrive at a test strategy and so on.

> **Note**
>
> *Figure 1.16* depicts the existence of multiple models. There may be several other perspectives, for example, a customer experience model and an information security model, which are not depicted.

Care should be taken to retain focus on solving the business problem at hand at all times. Teams will be better served if they expend the same amount of effort modeling business logic as the technical aspects of the solution. To keep accidental complexity in check, it will be best to isolate the infrastructure aspects of the solution from this model. These models can take several forms, including conversations, whiteboard sessions, documentation, diagrams, tests, and other forms of architecture fitness functions. It is also important to note that this is *not* a one-time activity. As the business evolves, the domain model and the solution will need to keep up. This can only be achieved through close collaboration between the domain experts and the developers at all times.

Scope of domain models and the bounded context

When creating domain models, one of the dilemmas is in deciding how to restrict the scope of these models. You can attempt to create a single domain model that acts as a solution for the entire problem. On the other hand, we may go the route of creating extremely fine-grained models that cannot exist meaningfully without having a strong dependency on others. There are pros and cons in going each way. Whatever be the case, each solution has a scope—bounds to which it is confined. This boundary is termed a **bounded context**.

There seems to exist a lot of confusion between the terms subdomains and bounded contexts. What is the difference? It turns out that subdomains are problem space concepts, whereas bounded contexts are solution space concepts. This is best explained through the use of an example. Let's consider the example of a fictitious Acme bank that provides two products: credit cards and retail bank. This may decompose to the following subdomains depicted here:

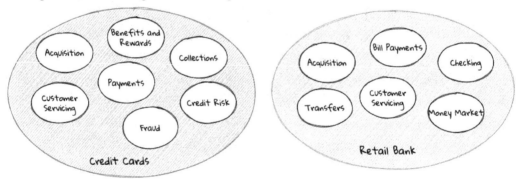

Figure 1.17 – Banking subdomains at Acme bank

When creating a solution for the problem, many possible solution options exist. We have depicted a few options here:

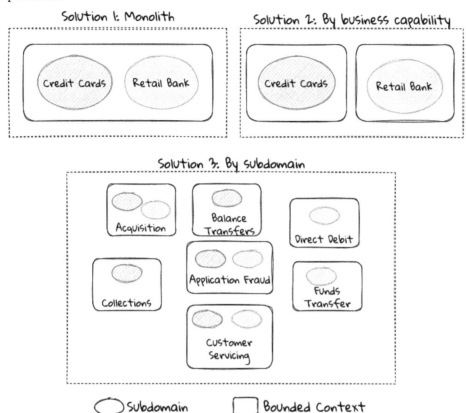

Figure 1.18 – Bounded context options at Acme bank

These are just a few examples of decomposition patterns to create bounded contexts. The exact set of patterns you may choose to use may vary depending on currently prevailing realities, such as the following:

- Current organizational structures
- Domain experts' responsibilities
- Key activities and pivotal events
- Existing applications

> **Note**
>
> **Conway's law** asserts that organizations are constrained to produce application designs that are copies of their communication structures. Your current organizational structures may not be optimally aligned to your desired solution approach. The inverse Conway maneuver may be applied to achieve isomorphism with the business architecture. Whatever the method used to decompose a problem into a set of bounded contexts, care should be taken to make sure that the coupling between them is kept as low as possible.

While bounded contexts ideally need to be as independent as possible, they may still need to communicate with each other. When using DDD, the system as a whole can be represented as a set of bounded contexts that have relationships with each other. These relationships define how these bounded contexts can integrate with each other and are called **context maps**. A sample context map is shown here:

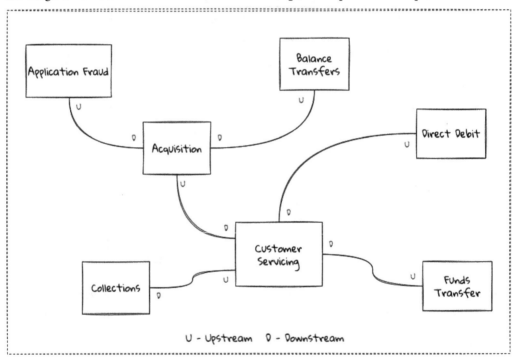

Figure 1.19 – Sample context map for Acme bank

The context map shows the bounded contexts and the relationship between them. These relationships can be a lot more nuanced than what is depicted here. We will discuss context maps and communication patterns in *Chapter 9, Integrating with External Systems*.

We have now covered a catalog of concepts that are core to the strategic design tenets of DDD. Let's look at some tools that can help expedite this process.

In subsequent chapters, we will reinforce all the concepts introduced here in a lot more detail.

In the next section, we will look at why the ideas of DDD, introduced all those years ago, are still very relevant. We will see why, if anything, they are becoming even more relevant now than ever.

Implementing the solution using tactical design

In the previous section, we saw how we can arrive at a shared understanding of the problem using strategic design tools. We need to use this understanding to create a solution. DDD's tactical design aspects, tools, and techniques help translate this understanding into working software. Let's look at these aspects in detail. In *Part 2* of the book, we will apply these to solve a real-world problem.

It is convenient to think of the tactical design aspects, as depicted in this figure:

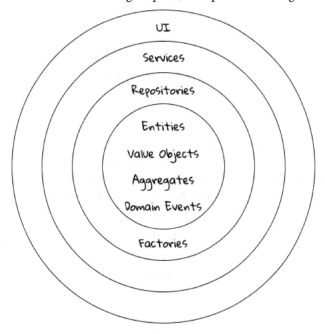

Figure 1.20 – The elements of DDD's tactical design

Let's look at the definitions of these elements.

Value objects

Value objects are immutable objects that encapsulate the data and behavior of one or more related attributes. It may be convenient to think of value objects as named primitives. For example, consider a `MonetaryAmount` value object. A simple implementation can contain two attributes—an amount and a currency code. This allows encapsulation of behavior, such as adding two `MonetaryAmount` objects safely, as shown here:

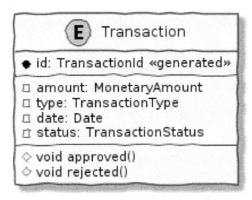

Figure 1.21 – A simple MonetaryAmount value object

The effective use of value objects helps protect from the primitive obsession with anti-patterns while increasing clarity. It also allows composing higher-level abstractions using one or more valuable objects. It is important to note that value objects do not have the notion of identity. That is, two values with the same value are treated equally. So, two `MonetaryAmount` objects with the same amount and currency code will be considered equal. Also, it is important to make value objects immutable. A need to change any of the attributes should result in the creation of a new attribute.

It is easy to dismiss value objects as a mere engineering technique, but the consequences of (not) using them can be far-reaching. In the `MonetaryAmount` example, it is possible for the *amount* and *currency code* to exist as independent attributes. However, the use of `MonetaryAmount` enforces the notion of the *ubiquitous language*. Hence, we recommend the use of value objects as a default instead of using primitives.

Critics may be quick to point out problems such as class explosion and performance issues. But in our experience, the benefits usually outweigh the costs. But it may be necessary to re-examine this approach if problems occur.

Entities

An entity is an object with a *unique identity* and *encapsulates* the data and behavior of its attributes. It may be convenient to view entities as a collection of other entities and value objects that need to be grouped together. A very simple example of an entity is shown here:

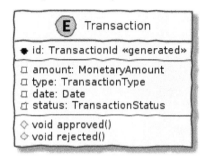

Figure 1.22 – A simple depiction of a Transaction entity

In contrast to a value object, entities have the notion of a unique identifier. This means that two `Transaction` entities with the same underlying values but a different **identifier** (**id**) value will be considered different. On the other hand, two entity instances with the same value for the identifier are considered equal. Furthermore, unlike value objects, entities are mutable. That is, their attributes can and will change over time.

The concept of value objects and entities depends on the context within which they are used. In an order management system, the *address* may be implemented as a value object in the *e-commerce* bounded context, whereas it may be needed to be implemented as an entity in the *order fulfillment* bounded context.

Important Note

It is common to collectively refer to entities and value objects as *domain objects*.

Aggregates

As seen previously, entities are hierarchical in that they can be composed of one more child. Fundamentally, an aggregate has the following qualities:

- Is an entity usually composed of other child entities and value objects

- Encapsulates access to child entities by exposing behavior (usually referred to as *commands*)

- Is a boundary that is used to enforce business invariants (rules) consistently

- Is an entry point to get things done within a bounded context

Consider the example of a `CheckingAccount` aggregate:

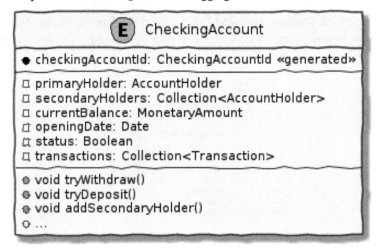

Figure 1.23 – A simple depiction of a CheckingAccount aggregate

Note how `CheckingAccount` is composed of the `AccountHolder` and `Transaction` entities, among other things. In this example, let's assume that the overdraft feature (ability to hold a negative account balance) is only available for **high-net-worth individuals (HNI)**. Any attempt to change `currentBalance` needs to occur in the form of a unique `Transaction` for audit purposes—irrespective of its outcome. For this reason, the `CheckingAccount` aggregate makes use of the `Transaction` entity. Although `Transaction` has `approve` and `reject` methods as part of its interface, only the aggregate has access to these methods. In this way, the aggregate enforces the business invariant while maintaining high levels of encapsulation. A potential implementation of the `tryWithdraw` method is shown here:

```
class CheckingAccount {
    private AccountHolder primaryHolder;                                    ❶
    private Collection<Transaction> transactions;                           ❶
    private MonetaryAmount currentBalance;                                  ❶
    // Other code omitted for brevity

    void tryWithdraw(MonetaryAmount amount) {                               ❷
        MonetaryAmount newBalance = currentBalance.subtract(amount);
        Transaction transaction = add(Transaction.withdrawal(id, amount));
        if (primaryHolder.isNotHNI() && newBalance.isOverdrawn()) { ❸
            transaction.rejected();
        } else {
            transaction.approved();
            currentBalance = newBalance;
        }
    }
}
```

1. The `CheckingAccount` aggregate is composed of child entities and value objects.

2. The `tryWithdraw` method acts as a consistency boundary for the operation. Irrespective of the outcome (approved or rejected), the system will remain in a consistent state. In other words, the `currentBalance` can change only within the confines of the `CheckingAccount` aggregate.

3. The aggregate enforces the appropriate business invariant (rule) to allow overdrafts only for HNIs.

> **Important Note**
>
> Aggregates are also referred to as aggregate roots, that is, the object that is at the root of the entity hierarchy. We use these terms synonymously in this book.

Domain events

As mentioned previously, aggregates dictate how and when state changes occur. Other parts of the system may be interested in knowing about the occurrence of changes that are significant to the business, for example, an order is placed or payment is received. *Domain events* are the means to convey that something significant to the business has occurred. It is important to differentiate between system events and domain events. For example, in the context of a retail bank, a *row was saved* in the database or a *server ran out of disk space* may classify as system events, whereas a *deposit was made* to a checking account and *fraudulent activity was detected* on a transaction could be classified as domain events. In other words, domain events are things that *domain experts care about.*

It may be prudent to make use of domain events to reduce the amount of coupling between bounded contexts, making it a critical building block of DDD.

Repositories

Most businesses require the durability of data. For this reason, the aggregate state needs to be persisted and retrieved when needed. Repositories are objects that enable persisting and loading *aggregate* instances. This is well documented in Martin Fowler's *Patterns of Enterprise Application Architecture* book as part of the *repository* (`https://martinfowler.com/eaaCatalog/repository.html`) pattern. It is pertinent to note that we are referring to aggregate repositories here, not just any entity repository. The singular purpose of this repository is to load a *single instance* of an aggregate using its identifier. It is important to note that this repository does not support finding aggregate instances using any other means. This is because business operations happen as part of manipulating a single instance of the aggregate within its bounded context.

Factories

In order to work with aggregates and value objects, instances of these need to be constructed. In simple cases, it might suffice to use a constructor to do so. However, aggregate and value object instances can become quite complex depending on the amount of state they encapsulate. In such cases, it may be prudent to consider delegating object construction responsibilities to a *factory* external to the aggregate/value object. We make use of the static factory method, builder, and dependency injection quite commonly in our day-to-day work. Joshua Bloch discusses several variations of this pattern in *Chapter 2, Where and How Does DDD Fit?*.

Services

When working within the confines of a single bounded context, the public interface (commands) of the aggregate provides a natural API. However, more complex business operations may require interacting with multiple bounded contexts and aggregates. In other words, we may find ourselves in situations where certain business operations do not fit naturally with any single aggregate. Even if interactions are limited to a single bounded context, there may be a need to expose that functionality in an implementation-neutral manner. In such cases, you may consider the use of objects called *services*. Services come in at least three flavors:

- **Domain services**: To enable coordinating operations among more than one aggregate – for example, transferring money between two checking accounts at a retail bank.

- **Infrastructure services**: To enable interactions with a utility that is not core to the business – for example, logging and sending emails at the retail bank.

- **Application services**: To enable coordination between domain services, infrastructure services, and other application services – for example, sending email notifications after a successful inter-account money transfer.

Services can also be stateful or stateless. It is best to allow aggregates to manage state, making use of repositories, while allowing services to coordinate and/or orchestrate business flows. In complex cases, there may be a need to manage the state of the flow itself. We will look at more concrete examples in *Part 2* of this book.

It may become tempting to implement business logic almost exclusively using services—inadvertently leading to the anemic domain model anti-pattern (`https://martinfowler.com/bliki/AnemicDomainModel.html`). It is worthwhile striving to encapsulate business logic within the confines of aggregates as a default.

Why is DDD relevant? Why now?

He who has a why to live for can bear almost any how.

— *Friedrich Nietzsche*

In a lot of ways, DDD was way ahead of its time when Eric Evans introduced the concepts and principles back in 2003. DDD seems to have gone from strength to strength. In this section, we will examine why DDD is even more relevant today than it was when Eric Evans wrote his book on the subject way back in 2003.

Rise of open source

Eric Evans, during his keynote address at the Explore DDD conference in 2017, lamented how difficult it was to implement even the simplest concepts, such as immutability in value objects, when his book had released. In contrast, though, nowadays, it's simply a matter of importing mature, well-documented, tested libraries, such as Project Lombok (`https://projectlombok.org/`) or Immutables (`https://immutables.github.io/`), to be productive, literally in a matter of minutes. To say that open source software has revolutionized the software industry would be an understatement! At the time of writing, the public Maven repository (`https://mvnrepository.com`) indexes no less than a staggering **18.3 million artifacts** in a large assortment of popular categories ranging from databases and language runtimes to test frameworks, and many, many more, as shown in the following chart:

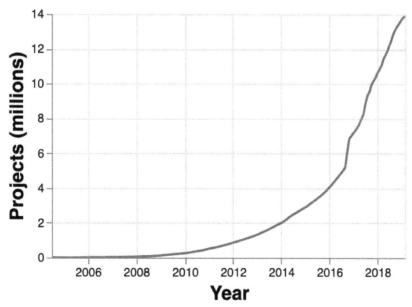

Figure 1.24 – Open source Java over the years (source: `https://mvnrepository.com/`)

Java stalwarts such as the Spring Framework and more recent innovations such as Spring Boot and Quarkus make it a no-brainer to create production-grade applications literally in a matter of minutes. Furthermore, frameworks such as Axon and Lagom, among others, make it relatively simple to implement advanced architecture patterns such are CQRS and event sourcing, which is very complementary to implementing DDD-based solutions.

Advances in technology

DDD by no means is just about technology; it could not be completely agnostic to the choices available at the time. 2003 was the heyday of heavyweight and ceremony-heavy frameworks, such as **Java 2 Enterprise Edition (J2EE)**, **Enterprise JavaBeans (EJB)**, SQL databases, and **object relational mappers (ORMs)**—with not much choice beyond that when it came to enterprise tools and patterns to build complex software, at least out in the public domain. The software world has evolved and come a very long way from there. In fact, modern game-changers such as Ruby on Rails and the public cloud were just getting released. In contrast, though, we now have no shortage of application frameworks, NoSQL databases, and programmatic APIs to create infrastructure components with a lot more releasing with monotonous regularity.

All these innovations allow for rapid experimentation, continuous learning, and iteration at pace. These game-changing advances in technology have also coincided with the exponential rise of the internet and e-commerce as viable means to carry out successful businesses. In fact, the impact of the internet is so pervasive that it is almost inconceivable to launch businesses without a digital component being an integral component. Finally, the consumerization and wide-scale penetration of smartphones, IoT devices, and social media have meant that data is being produced at rates inconceivable as recent as a decade ago. This means that we are building for and solving the most complicated problems by several orders of magnitude.

Rise of distributed computing

There was a time when building large monoliths was very much the default. But an exponential rise in computing technology, the public cloud (IaaS, PaaS, SaaS, and FaaS), big data storage, and processing volumes, which has coincided with an arguable slowdown in the ability to continue creating faster CPUs, has meant a turn toward more decentralized methods of solving problems.

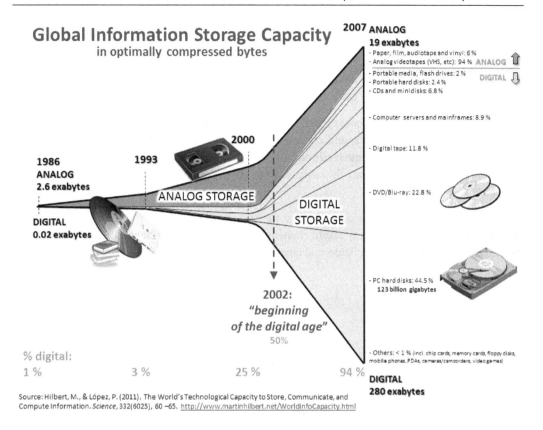

Figure 1.25 – Global information storage capacity

DDD, with its emphasis on dealing with complexity by breaking unwieldy monoliths into more manageable units in the form of subdomains and bounded contexts, fits naturally into this style of programming. Hence, it is no surprise to see a renewed interest in adopting DDD principles and techniques when crafting modern solutions. To quote Eric Evans, it is no surprise that DDD is even more relevant now than when it was originally conceived!

Summary

In this chapter, we examined some common reasons why software projects fail. We saw how inaccurate or misinterpreted requirements, architecture (or the lack thereof), and excessive technical debt can get in the way of meeting business goals and success.

We looked at the basic building blocks of DDD, such as domains, subdomains, ubiquitous language, domain models, bounded contexts, and context maps. We also examined why the principles and techniques of DDD are still very much relevant in the modern age of microservices and serverless. You should now be able to appreciate the basic terms of DDD and understand why it is important in today's context.

In the next chapter, we will take a closer look at the real-world mechanics of DDD. We will delve deeper into the strategic and tactical design elements of DDD and look at how using these can help form the basis for better communication and create more robust designs.

Further reading

Title	Author	Location
Pulse of the Profession – 2017	PMI	`https://www.pmi.org/-/media/pmi/documents/public/pdf/learning/thought-leadership/pulse/pulse-of-the-profession-2017.pdf`
Pulse of the Profession – 2020	PMI	`https://www.pmi.org/learning/library/forging-future-focused-culture-11908`
Project success: Definitions and Measurement Techniques	PMI	`https://www.pmi.org/learning/library/project-success-definitions-measurement-techniques-5460`
Project success: Definitions and Measurement Techniques	JK Pinto, DP Slevin	`https://www.pmi.org/learning/library/project-success-definitions-measurement-techniques-5460`
Analysis Paralysis	Ward Cunningham	`https://proxy.c2.com/cgi/wiki?AnalysisParalysis`
Big Design Upfront	Ward Cunningham	`https://wiki.c2.com/?BigDesignUpFront`
Enterprise Modeling Anti-Patterns	Scott W. Ambler	`http://agilemodeling.com/essays/enterpriseModelingAntiPatterns.htm`
A Project Manager's Guide To 42 Agile Methodologies	Henny Portman	`https://thedigitalprojectmanager.com/agile-methodologies`

Title	Author	Location
Domain-Driven Design Even More Relevant Now	Eric Evans	`https://www.youtube.com/watch?v=kIKwPNKXaLU`
Introducing Deliberate Discovery	Dan North	`https://dannorth.net/2010/08/30/introducing-deliberate-discovery/`
No Silver Bullet—Essence and Accident in Software Engineering	Fred Brooks	`http://faculty.salisbury.edu/~xswang/Research/Papers/SERelated/no-silver-bullet.pdf`
Mastering Non-Functional Requirements	Sameer Paradkar	`https://www.packtpub.com/product/mastering-non-functional-requirements/9781788299237`
Big Ball Of Mud	Brian Foote & Joseph Yoder	`http://www.laputan.org/mud/`
The Forgotten Layer of the Test Automation Pyramid	Mike Cohn	`https://www.mountaingoatsoftware.com/blog/the-forgotten-layer-of-the-test-automation-pyramid`
Tech debt: Reclaiming tech equity	Vishal Dalal et al.	`https://www.mckinsey.com/business-functions/mckinsey-digital/our-insights/tech-debt-reclaiming-tech-equity`
Is High-Quality Software Worth the Cost?	Martin Fowler	`https://martinfowler.com/articles/is-quality-worth-cost.html#WeAreUsedToATrade-offBetweenQualityAndCost`

2

Where and How Does DDD Fit?

"We won't be distracted by comparison if we are captivated with purpose."

– Bob Goff

Software architecture refers to the fundamental structures of a software system and the discipline of creating such structures and systems. Over the years, we have accumulated a series of architecture styles and programming paradigms to help us deal with system complexity.

In this chapter, we will examine how **Domain-Driven Design** (DDD) can be applied in a manner that is complementary to these architecture styles and programming paradigms. We will also look at how/where it fits in the overall scheme of things when crafting a software solution.

In this chapter, we will cover the following topics:

- Architecture styles

- Programming paradigms

- Which paradigm should you choose?

By the end of this chapter, you will have gained an appreciation of a variety of architecture styles and programming paradigms, along with some pitfalls to watch out for when applying them. You will also understand the role that DDD plays in augmenting each of these.

Architecture styles

Domain-driven design presents a set of architecture tenets in the form of strategic and tactical design elements. This enables you to decompose large, potentially unwieldy business subdomains into well-factored, independent bounded contexts.

One of the great advantages of DDD is that it does not require the use of any specific architecture. However, the software industry has been using a plethora of architecture styles over the last few years. Let's look at how DDD can be used in conjunction with a set of popular architecture styles to arrive at better solutions.

Layered architecture

Layered architecture is one of the most common architecture styles, where the solution is typically organized into four broad categories: **presentation**, **application**, **domain**, and **persistence**. Each of the layers provides a solution to a particular concern it represents, as shown here:

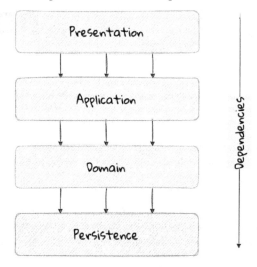

Figure 2.1 – The essence of layered architecture

The main idea behind layered architecture is a separation of concerns – where the dependencies between layers are unidirectional (from top to bottom). For example, the domain layer can depend on the persistence layer, not the other way round. In addition, any given layer typically accesses the layer immediately beneath it without bypassing layers in between. For example, the presentation layer may access the domain layer only through the application layer.

This structure enables looser coupling between layers and allows them to evolve independently of each other. The idea of layered architecture fits very well with DDD's tactical design elements, as depicted here:

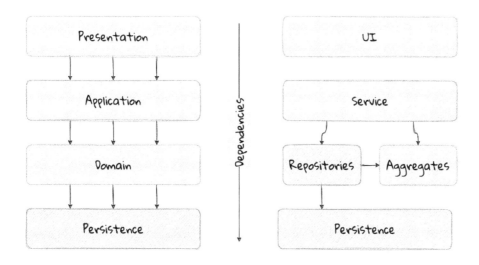

Figure 2.2 – Layered architecture mapped to DDD's tactical design elements

DDD actively promotes the use of a layered architecture, primarily because it makes it possible to focus on the domain layer in isolation from other concerns, such as how information gets displayed, how end-to-end flows are managed, and how data is stored and retrieved. From that perspective, solutions that apply DDD naturally tend to be layered as well.

Notable variations

A variation of layered architecture was invented by Alistair Cockburn, which he originally called hexagonal architecture (https://alistair.cockburn.us/hexagonal-architecture/), alternatively called ports and adapters architecture. The idea behind this style was to avoid inadvertent dependencies between layers (which could occur in layered architecture), specifically between the core of a system and the peripheral layers.

The main idea here is to make use of interfaces (**ports**) exclusively within the core to enable modern drivers, such as testing and looser coupling. This allows the core to be developed and evolved independently of the non-core parts and the external dependencies. Integration with real-world components such as databases, filesystems, and web services is achieved through concrete implementations of the ports (**adapters**). The use of interfaces within the core enables much easier testing of the core in isolation from the rest of the system, using mocks and stubs. It is also common to use dependency injection frameworks to dynamically swap out implementations of these interfaces when working with a real system in an end-to-end environment. A visual representation of hexagonal architecture is shown here:

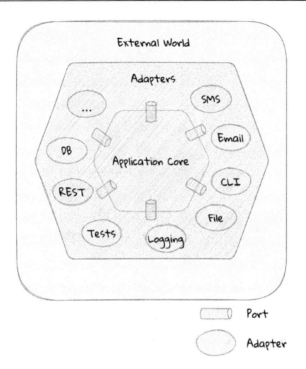

Figure 2.3 – Hexagonal architecture

It turns out that the use of the term *hexagon* in this context was purely for visual purposes– not to limit the system to exactly six types of ports.

Similar to the hexagonal architecture, onion architecture (`https://jeffreypalermo.com/2008/07/the-onion-architecture-part-1/`), conceived by Jeffrey Palermo, is based on creating an application based on an independent object model within the core that can be compiled and run separately from the outer layers. This is done by defining interfaces (*ports* in hexagonal architecture) in the core and implementing them (*adapters* in hexagonal architecture) in the outer layers. From our perspective, the hexagonal and onion architecture styles have no perceptible differences that we could identify.

A visual representation of onion architecture is shown here:

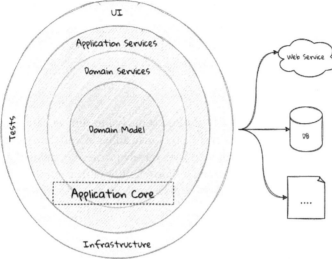

Figure 2.4 – Onion architecture

Yet another variation of layered architecture, popularized by Robert C. Martin (known endearingly as Uncle Bob), is clean architecture. This is based on adhering to the SOLID principles (https://blog.cleancoder.com/uncle-bob/2020/10/18/Solid-Relevance.html), also conceived by him. The fundamental message here (just like in the case of hexagonal and onion architecture) is to avoid dependencies between the core – the one that houses business logic – and other layers that tend to be volatile (such as frameworks, third-party libraries, UIs, and databases).

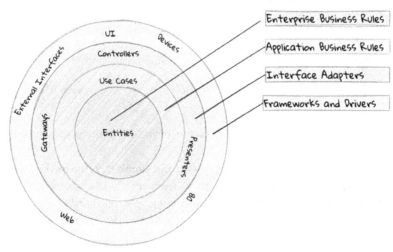

Figure 2.5 – Clean architecture

All these architecture styles are synergistic with DDD's idea of developing a domain model for the core subdomain (and, by extension, its bounded context) independently of the rest of the system.

While each of these architecture styles provides additional guidance in terms of how to structure a layered architecture, any architecture approach we choose comes with its set of trade-offs and limitations you will need to be cognizant of. We will discuss some of these considerations in the next sub-sections.

The layer cake anti-pattern

Sticking to a fixed set of layers provides a level of isolation, but in simpler cases, it may prove to be overkill without adding any perceptible benefit, other than adherence to the agreed-on architecture guidelines. In the layer cake anti-pattern, each layer merely proxies the call to the layer beneath it without adding any value. The following example illustrates this fairly common scenario:

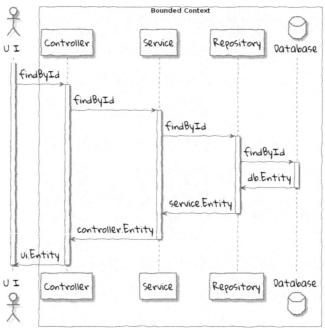

Figure 2.6 – An example of the layer cake anti-pattern to find an entity representation by ID

Here, the findById method is replicated in every layer and simply calls the method with the same name in the layer below, with no additional logic. This introduces a level of accidental complexity to the solution. Some amount of redundancy in the layering may be unavoidable for the purposes of standardization. It may be best to reexamine the layering guidelines if the *layer cake* occurs prominently in the code base.

Anemic translation

Another variation of the layer cake we see commonly is one where layers refuse to share input and output types in the name of higher isolation and looser coupling. This makes it necessary to perform translations at the boundary of each layer. If the objects being translated are more or less structurally identical, we have an **anemic translation**. Let's look at a variation of the `findById` example we discussed previously.

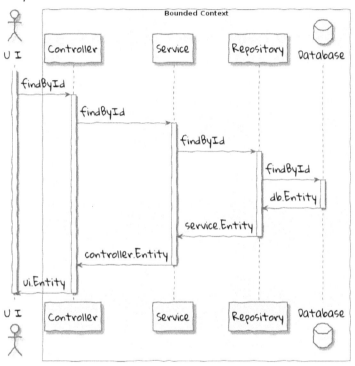

Figure 2.7 – An example of the anemic translation anti-pattern

In this case, each layer defines an `Entity` type of its own, requiring a translation between types at each layer. To make matters worse, the structure of the `Entity` type may have seemingly minor variations (for example, `lastName` being referred to as `surname`). While such translations may be necessary across bounded contexts, teams should strive to avoid the need for variations in names and structures of the same concept within a single bounded context. The intentional use of *ubiquitous language* helps avoid such scenarios.

Layer bypass

When working with a layered architecture, it is reasonable to start by being strict about layers only interacting with the layers immediately beneath them. As we saw earlier, such rigid enforcements may lead to an intolerable degree of accidental complexity, especially when applied generically to a large number of use cases. In such scenarios, it may be worth considering consciously allowing one or more layers to be bypassed.

For example, the `controller` layer may be allowed to work directly with the `repository` without using the `service` layer. In many cases, we have found it useful to use a separate set of rules for *commands* versus *queries* as a starting point.

This can be a slippery slope. To continue maintaining a level of sanity, teams should consider the use of a lightweight architecture governance tool such as **ArchUnit** (`https://www.archunit.org/`) to make agreements explicit and provide quick feedback. A simple example of how to use ArchUnit for this purpose is shown here:

```
class LayeredArchitectureTests {
    @ArchTest
    static final ArchRule layer_dependencies_are_respected_with_exception =
layeredArchitecture()

            .layer("Controllers").definedBy("..controller..")
            .layer("Services").definedBy("..service..")
            .layer("Domain").definedBy("..domain..")
            .layer("Repository").definedBy("..repository..")

            .whereLayer("Controllers").mayNotBeAccessedByAnyLayer()
            .whereLayer("Services").mayOnlyBeAccessedByLayers("Controllers")
            .whereLayer("Domain").mayOnlyBeAccessedByLayers("Services",
    "Repository", "Controllers")
            .whereLayer("Repository")
                .mayOnlyBeAccessedByLayers("Services", "Controllers"); ❶
}
```

The `Repository` layer can be accessed by both the Services and Controllers layers – effectively allowing Controllers to bypass the use of the Services layer.

Vertical slice architecture

Layered architecture and its variants provide reasonably good guidance on how to structure complex applications. Vertical slice architecture, championed by Jimmy Boggard, recognizes that it may be too rigid to adopt a standard layering strategy for all use cases across an entire application.

Furthermore, it is important to note that business value cannot be derived by implementing any of these horizontal layers in isolation. Doing so will only result in unusable inventory and lots of unnecessary context switching until all these layers are connected. Therefore, vertical slice architecture proposes minimizing coupling between slices and maximizing coupling in a slice (`https://jimmybogard.com/vertical-slice-architecture/`), as shown here:

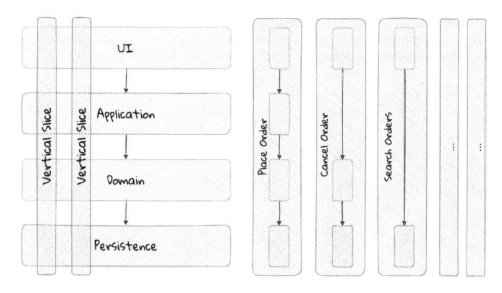

Figure 2.8 – Vertical slice architecture

In this example, **Place Order** might require us to coordinate with other components through the application layer and apply complex business invariants, while operating within the purview of an ACID transaction. Similarly, **Cancel Order** might require applying business invariants within an ACID transaction without any additional coordination – obviating the need for the application layer, in this case. However, **Search Orders** might require us to simply fetch existing data from a query-optimized view. This style makes use of a horses-for-courses approach to layering that may help alleviate some anti-patterns listed previously when implementing a plain vanilla layered architecture.

Considerations

Vertical slice architecture affords a lot of flexibility when implementing a solution – taking into consideration the specific needs of the use case being implemented. However, without some level of governance, this may quickly devolve into a big ball of mud, with layering decisions being made seemingly arbitrarily, based on personal preferences and experiences (or lack thereof). As a sensible default, you may want to consider using a distinct layering strategy for commands and queries. Beyond that, non-functional requirements may dictate how you may need to deviate from here. For example, you may need to bypass layers to meet performance SLAs for certain use cases.

When used pragmatically, vertical slice architecture does enable you to apply DDD very effectively within each or a group of related vertical slices – allowing them to be treated as bounded contexts. Two possibilities using the place order and cancel order examples are shown here:

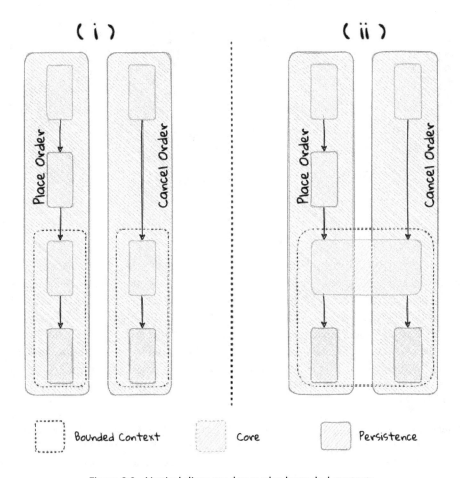

Figure 2.9 – Vertical slices used to evolve bounded contexts

In the (i) example in the preceding diagram, place order and cancel order each use a distinct domain model, whereas in the (ii) example, both use cases share a common domain model and, by extension, become part of the same bounded context. This does pave the way to slice functionality when looking to adopt the serverless architecture along use case boundaries.

Service-oriented architecture (SOA)

Service-Oriented Architecture (**SOA**) is an architecture style where software components expose (potentially) reusable functionality over standardized interfaces. The use of standardized interfaces (such as SOAP, REST, and gRPC, to name a few) enables easier interoperability when integrating heterogeneous solutions, as shown here:

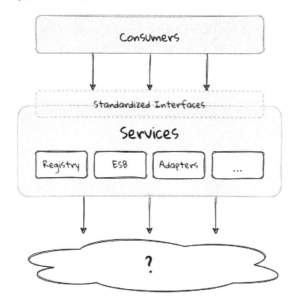

Figure 2.10 – SOA – exposing reusable functionality over standard interfaces

Previously, the use of non-standard, proprietary interfaces made this kind of integration a lot more challenging. For example, a retail bank may expose inter-account transfer functionality in the form of SOAP web services. While SOA prescribes exposing functionality over standardized interfaces, the focus is more on integrating heterogeneous applications than implementing them.

Considerations

At one of the banks we worked at, we exposed a set of over 500 service interfaces over SOAP. Under the hood, we implemented these services using EJB 2.x (a combination of stateless session beans and message-driven beans), hosted on a commercial J2EE application server, which also did double duty as an **Enterprise Service Bus** (**ESB**). These services delegated most, if not all, of the logic to a set of underlying stored procedures within a single monolithic Oracle database, using a canonical data model for the entire enterprise! To the outside world, these services were *location-transparent*, stateless, *composable*, and *discoverable*. Indeed, we advertised this implementation as an example of SOA, and it would be hard to argue that it was not.

This suite of services had evolved organically over the years, with no explicit boundaries, concepts from various parts of the organization, and generations of people mixed in, each adding their own interpretation of how business functionality needed to be implemented. In essence, the implementation resembled the dreaded big ball of mud, which was extremely hard to enhance and maintain.

The intentions behind SOA are noble. However, the promises of *reuse* and *loose coupling* are hard to achieve in practice, given the lack of concrete implementation guidance on component granularity. It is also true that SOA means many things (`https://martinfowler.com/bliki/ServiceOrientedAmbiguity.html`) to different people. This ambiguity leads to most SOA implementations becoming complex, unmaintainable monoliths, centered around technology components such as a service bus, a persistence store, or both. This is where using DDD to solve a complex problem by breaking it down into subdomains and bounded contexts can be invaluable.

Microservices architecture

In the last decade or so, microservices have gained quite a lot of popularity, with lots of organizations wanting to adopt this style of architecture. In a lot of ways, microservices are an extension of SOA – one where much emphasis is placed on creating focused components that deal with doing a limited number of things and doing them right. Sam Newman, the author of the *Building Microservices* book, defines microservices as *small-sized*, independently deployable components that maintain their own state and are *modeled around a business domain*. This affords benefits such as adopting a horses-for-courses approach when modeling solutions, limiting the blast radius, improved productivity and speed, autonomous cross-functional teams, and so on.

Microservices usually exist as a collective, working collaboratively to achieve desired business outcomes, as depicted here:

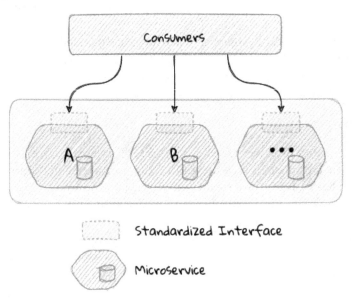

Figure 2.11 – A microservices ecosystem

As we can see, SOA and microservices are very similar from the perspective of the consumers, in that they access functionality through a set of standardized interfaces. The microservices approach is an evolution of SOA in that the focus now is on building smaller, self-sufficient, independently deployable components with the intent of avoiding single points of failure (such as an enterprise database or service bus), which was fairly common with a number of SOA-based implementations.

Considerations

While microservices have definitely helped, there still exists quite a lot of ambiguity when it comes to answering how big or small a microservice should be (`https://martinfowler.com/articles/microservices.html#HowBigIsAMicroservice`). Indeed, a lot of teams seem to struggle to get this balance right, resulting in a distributed monolith (`https://www.infoq.com/news/2016/02/services-distributed-monolith/`), which, in a lot of ways, can be much worse than even a single-process monolith from the SOA days. Again, applying the strategic design concepts of DDD can help create independent, loosely coupled components, making it an ideal companion for the microservices style of architecture.

Event-driven architecture (EDA)

Irrespective of the granularity of components (monolith, microservices, or something in between), most non-trivial solutions have a boundary, beyond which there may be a need to communicate with external system(s). This communication usually happens through the exchange of messages between systems, causing them to become coupled with each other. Coupling comes in two broad flavors: *afferent* – who depends on you, and *efferent* – who you depend on. Excessive amounts of efferent coupling can make systems very brittle and hard to work with.

Event-driven systems enable authoring solutions that have a relatively low amount of efferent coupling by emitting events when they attain a certain state, without caring about who consumes those events. In this regard, it is important to differentiate between message-driven and event-driven systems, as mentioned in the *Reactive Manifesto*:

> *"A message is an item of data that is sent to a specific destination. An event is a signal emitted by a component upon reaching a given state. In a message-driven system addressable recipients await the arrival of messages and react to them, otherwise lying dormant. In an event-driven system notification listeners are attached to the sources of events such that they are invoked when the event is emitted. This means that an event-driven system focuses on addressable event sources while a message-driven system concentrates on addressable recipients."*
>
> *– The Reactive Manifesto*

In simpler terms, event-driven systems do not care who the downstream consumers are, whereas in a message-driven system, that may not necessarily be true. When we say event-driven in the context of this book, we mean the former.

Typically, event-driven systems eliminate the need for point-to-point messaging with the ultimate consumers by making use of an intermediary infrastructure component, usually known as a message broker, event bus, and so on. This effectively reduces efferent coupling from n consumers to 1. There are a few variations on how event-driven systems can be implemented. In the context of publishing events, Martin Fowler talks about two broad styles (among other things), event notifications and event-carried state transfer, in his *What do you mean by "Event-Driven"?* article (`https://martinfowler.com/articles/201701-event-driven.html`).

Considerations

One of the main trade-offs when building an event-driven system is to decide the amount of state (payload) that should be embedded in each event. It may be prudent to consider embedding just enough state to indicate changes that occurred as a result of the emitted event to keep the various opposing forces, such as producer scaling, encapsulation, consumer complexity, and resiliency. We will discuss the related implications in more detail when we cover implementing events in *Chapter 5, Implementing Domain Logic*.

DDD is all about keeping complexity in check by creating these independent bounded contexts. However, *independent* does not mean *isolated*. Bounded contexts may still need to communicate with each other. One way to do that is through the use of a fundamental DDD building block – domain events. Event-driven architecture and DDD are thus complementary. It is typical to make use of an event-driven architecture to allow bounded contexts to communicate while continuing to loosely couple with each other.

Command Query Responsibility Segregation (CQRS)

In traditional applications, a single-domain, data/persistence model is used to handle all kinds of operations. With CQRS, we create distinct models to handle updates (commands) and inquiries. This is depicted in the following diagram:

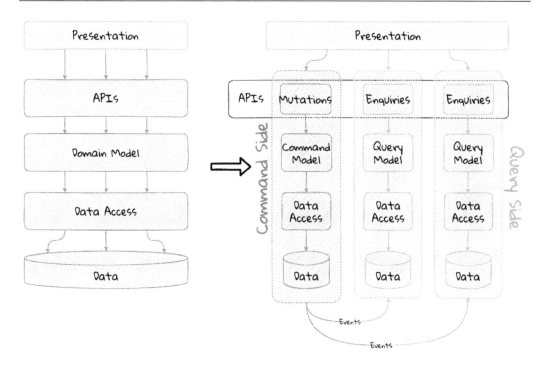

Figure 2.12 – Traditional versus CQRS architecture

> **Note**
>
> We depict multiple query models in the previous diagram because it is possible (but not necessary) to create more than one query model, depending on the kinds of query use cases that need to be supported.

For this to work predictably, the query model(s) need to be kept in sync with the write models (we will examine some of the techniques to do that in detail later).

Considerations

A traditional, single-model approach works well for simple, CRUD-style applications but starts to become unwieldy for more complex scenarios. We will discuss some of these scenarios in the next subsections.

Volume imbalance between reads and writes

In most systems, read operations often outnumber write operations by significant orders of magnitude. For example, consider the number of times a trader checks stock prices versus the number of times they actually transact (buy or sell stock trades). It is also usually true that write operations are the ones that make businesses money. Having a single model for both reads and writes in a system with a majority of read operations can overwhelm a system to an extent where write performance can become affected.

A need for multiple read representations

When working with relatively complex systems, it is not uncommon to require more than one representation of the same data. For example, when looking at personal health data, you may want to look at a daily, weekly, or monthly view. While these views can be computed on the fly from the *raw* data, each transformation (aggregation, summarization, and so on) adds to the cognitive load on a system. Often, it is not possible to predict ahead of time the nature of these requirements. By extension, it is not feasible to design a single canonical model that can provide answers to all these requirements. Creating domain models specifically designed to meet a focused set of requirements can be much easier.

Different security requirements

Managing the authorization of and access requirements to data/APIs when working with a single model can start to become cumbersome. For example, higher levels of security may be desirable for debit operations in comparison to balance enquiries. Having distinct models can considerably ease the complexity of designing fine-grained authorization controls.

More uniform distribution of complexity

Having a model dedicated to serving only command-side use cases means that they can now be focused toward solving a single concern. For query-side use cases, we create models as needed that are distinct from the command-side model. This helps spread complexity more uniformly over a larger surface area – as opposed to increasing the complexity on a single model that is used to serve all use cases. It is worth noting that the essence of DDD is mainly to work effectively with complex software systems, and CQRS fits well with this line of thinking.

> **Note**
> When working with a CQRS-based architecture, choosing the persistence mechanism for the command side is a key decision. When working in conjunction with event-driven architecture, you could choose to persist aggregates as a series of events (ordered in the sequence of their occurrence). This style of persistence is known as event sourcing. We will cover this in more detail in *Chapter 5, Implementing Domain Logic*, in the *Event-sourced aggregates* section.

Serverless architecture

Serverless architecture is an approach to software design that allows developers to build and run services without having to manage the underlying infrastructure. The advent of the AWS Lambda service has popularized this style of architecture, although several other services (such as S3 and DynamoDB for persistence, SNS for notifications, and SQS for message queuing) existed long before Lambda was launched. While AWS Lambda provides a compute solution in the form of **Functions as a Service (FaaS)**, these other services are just as essential, if not more, in order to benefit from a serverless paradigm.

In conventional DDD, bounded contexts are formed by grouping related operations around an aggregate, which then informs how the solution is deployed as a unit – usually within the confines of a single process. With a serverless paradigm, each operation (task) is expected to be deployed as an independent unit of its own. This requires that we look at how we model aggregates and bounded contexts differently – now centered around individual tasks or functions, as opposed to a group of related tasks.

Does that mean that the principles of DDD to arrive at a solution do not apply anymore? While a serverless paradigm introduces an additional dimension of having to treat fine-grained deployable units as first-class citizens in the modeling process, the overall process of applying DDD's strategic and tactical design continues to apply. We will examine this in more detail in *Chapter 11, Decomposing into Finer-Grained Components*, when we will refactor the solution we build throughout this book to employ a serverless approach.

The big ball of mud

Thus far, we have examined a catalog of named architecture styles along with their pitfalls and how applying DDD can help alleviate them. At the other extreme, we may encounter solutions that lack a perceivable architecture, infamously known as the *big ball of mud*:

> *"A BIG BALL OF MUD is haphazardly structured, sprawling, sloppy, duct-tape and baling wire, spaghetti code jungle. We've all seen them. These systems show unmistakable signs of unregulated growth, and repeated, expedient repair. Information is shared promiscuously among distant elements of the system, often to the point where nearly all the important information becomes global or duplicated. The overall structure of the system may never have been well-defined. If it was, it may have eroded beyond recognition. Programmers with a shred of architectural sensibility shun these quagmires. Only those who are unconcerned about architecture, and, perhaps, are comfortable with the inertia of the day-to-day chore of patching the holes in these failing dikes, are content to work on such systems."*
>
> *– Brian Foote and Joseph Yoder*

Although Foote and Yoder advise avoiding this style of architecture at all costs, software systems that resemble the big ball of mud continue to be a day-to-day inevitability for a lot of us. The strategic and tactical design elements of DDD provide a set of techniques to help deal with and recover from these near-hopeless situations in a pragmatic manner, without potentially having to adopt a big bang approach. Indeed, the focus of this book is to apply these principles to prevent or at least delay further devolution toward the big ball of mud.

Which architecture style should you use?

As we have seen, there are a variety of architecture styles you can lean on when crafting a software solution. A lot of these architecture styles share quite a few common tenets. It can become difficult to conform to any single architecture style. DDD, with its emphasis on breaking down complex business problems into subdomains and bounded contexts, enables the use of more than one approach across bounded contexts. We would like to make a special mention of vertical slice architecture because it places an emphasis on dividing functionality into specific business outcomes and, thus, more naturally adheres to DDD's ideas of subdomains and bounded contexts. In reality, you may find the need to extend and even deviate from pedantic definitions of architecture styles in order to meet real-world needs. But when we do make such compromises, it is important to do so *intentionally* and make it unambiguously clear why we are making such a decision (preferably using some lightweight mechanism, such as **ADRs** (`https://www.thoughtworks.com/de-de/radar/techniques/lightweight-architecture-decision-records`). This is important because it may become hard to justify this to others and even ourselves when we look at it in the future.

In this section, we have examined popular architecture styles and how we can amplify their effectiveness when used in conjunction with DDD. Now, let's look at how DDD can complement the use of existing programming paradigms.

Programming paradigms

The tactical elements of DDD introduce a specific vocabulary (aggregates, entities, value objects, repositories, services, factories, domain events, and so on) when arriving at a solution. At the end of the day, we need to translate these concepts into running software. Over the years, we have employed a variety of programming paradigms, including procedural, object-oriented, functional, and aspect-oriented. Is it possible to apply DDD in conjunction with one or more of these paradigms? In this section, we will explore how some common programming paradigms and techniques help us express tactical design elements in code.

Object-oriented programming

On the surface of it, DDD seems to simply replicate a set of OO terms and call them by different names. For example, the central concepts of tactical DDD such as aggregates, entities, and value objects could simply be referred to as objects in OO terms. Others such as services may not have a direct OO analog. So, how do you apply DDD in an OO world? Let's look at a simple example:

```
1   interface PasswordService {
2       String generateStrongPassword();
3       boolean isStrong(String password);
4       boolean isWeak(String password);
5   }
6
7   class PasswordClient {
8       private PasswordService service;
9
10      void register(String userEnteredPassword) {
11          if (service.isStrong(userEnteredPassword)) {
12              //...
13          }
14      }
15  }
16
```

OO purists will be quick to point out that `PasswordService` is procedural and that a `Password` class might be needed to encapsulate related behaviors. Similarly, DDD enthusiasts might point out that this is an anemic domain model implementation. An arguably better OO version might look something like the following:

```
1   class Password {
2       private final String password;
3
4       private Password(String password) {
5           this.password = password;
6       }
7
8       public boolean isStrong() { ... }
9       public boolean isWeak() { ... }
10      public static Password generateStrongPassword() { ... }
11      public static Password passwordFrom(String password) { ... }
12
13  }
14
15  interface PasswordService {
16      Password generateStrongPassword();
17      Password createPasswordFrom(String userEntered);
18  }
19
20  class PasswordClient {
21      private PasswordService service;
22
23      void register(String userEnteredPassword) {
24          Password password = service.createPasswordFrom(userEnteredPassword);
25          if (password.isStrong()) {
26              // ...
27          }
28      }
29  }
```

In this case, the `Password` class stops exposing its internals and exposes the idea of a strong or weak password in the form of behavior (the `isStrong` and `isWeak` methods). From an OO perspective, the second implementation is arguably superior. If so, shouldn't we be using the OO version at all times? As it turns out, the answer is nuanced and depends on what consumers desire and the ubiquitous language used in that context. If the concept of `Password` is in common usage within the domain, it perhaps warrants introducing such a concept in the implementation as well. If not, the first solution might suffice, even though it seems to violate OO principles of encapsulation.

Our default position is to apply good OO practices as a starting point. However, it is more important to mirror the language of the domain as opposed to applying OO in a dogmatic manner. So, we will be willing to compromise on OO purity if it appears unnatural to do so in that context. As mentioned earlier, clearly communicating the rationale for such decisions can go a long way.

Functional programming

Functions are a fundamental building block to code organization that exists in all higher-order programming languages. Functional programming is a programming paradigm where programs are constructed by applying and composing functions. This is in contrast to imperative programming, which uses statements to change a program's state. The most significant differences stem from the fact that functional programming avoids side effects, which are common in imperative programming. Pure functional programming completely prevents side effects and forces immutability. Embracing a functional style when designing a domain model to be more declarative expresses intent a lot more clearly while remaining terse. It also allows us to keep the complexity in check by enabling us to compose more complex concepts by using simpler ones. The functional implementation allows us to use a language closer to the problem domain, while having the added benefit of also being terse. Consider a simple example where we need to find the item with the least inventory across all our warehouses using a functional style, as shown here:

```
 1  class Functional {
 2      public static Optional<Item> scarcestItem(Warehouse... warehouses) {
 3          return Stream.of(warehouses)
 4                  .flatMap(Warehouse::items)
 5                  .collect(groupingBy(Item::name, summingInt(Item::quantity)))
 6                  .entrySet().stream()
 7                  .map(Item::new)
 8                  .min(comparing(Item::quantity));
 9      }
10  }
```

The imperative style shown here does get the job done but is arguably a lot more verbose and harder to follow, sometimes even for technical team members!

Here is an imperative example:

```
 1  class Imperative {
 2      public static Optional<Item> scarcestItem(Warehouse... warehouses) {
 3          Collection<Item> allItems = new ArrayList<>();
 4          for (Warehouse warehouse : warehouses) {
 5              allItems.addAll(warehouse.getItems());
 6          }
 7          Map<String, Integer> itemNamesByQuantity = new HashMap<>();
 8          for (Item item : allItems) {
 9              final String name = item.name();
10              final int quantity = item.quantity();
11              if (itemNamesByQuantity.containsKey(name)) {
12                  itemNamesByQuantity.put(name, itemNamesByQuantity.get(name)
13                                          + quantity);
14              } else {
15                  itemNamesByQuantity.put(name, quantity);
16              }
17          }
18          final Map.Entry<String, Integer> min =
19              Collections.min(itemNamesByQuantity.entrySet(),
20  Map.Entry.comparingByValue());
21          return min != null ? Optional.of(new Item(min)) : Optional.empty();
22
23      }
    }
```

From a DDD perspective, this yields a few benefits:

- **Increased collaboration with domain experts** because the declarative style allows placing a bigger focus on the what, rather than the how. This makes it a lot less intimidating to technical and non-technical stakeholders alike to work with on an ongoing basis.

- **Better testability** because the use of pure functions (those that are side-effect-free) makes it easier to create data-driven tests. This has also afforded us an additional benefit of less mocking/stubbing. These characteristics make tests that are a lot easier to maintain and rationalize. This has the benefit of allowing even technical team members to visualize corner cases a lot earlier in the process.

Which paradigm should you choose?

DDD simply states that you should build your software around a domain model that represents the actual problem that the software is trying to solve. When encountering complex real-life problems, we often find it hard to conform to any single paradigm across the board. Looking to use a one-size-fits-all approach may work to our detriment. Our experience indicates that we will need to make use of a variety of techniques in order to solve the problem at hand elegantly. Java is inherently an OO language, but with the advent of Java 8, it has started to embrace a variety of functional constructs as well. This allows us to make use of a multitude of techniques to create elegant solutions. The most important thing is to agree on a ubiquitous language and allow it to guide the approach taken. It also largely depends on the talent and experience you have at your disposal. Making use of a style that is foreign to the majority of a team will likely prove counterproductive. Although we haven't covered the procedural paradigm here in this chapter, there may be occasions where it might be the best solution given the current situation. As long as we are intentional about areas where we deviate from the accepted norm for a particular programming paradigm, we should be in a reasonably good place.

Summary

In this chapter, we covered a series of commonly used architecture patterns and how we can practice DDD when working with them. We also looked at common pitfalls and gotchas that we may need to be cognizant of when using these architectures. We also looked at popular programming paradigms and their influence on the tactical elements of DDD.

Additionally, you should have an appreciation of the various architecture styles that you need to employ when coming up with a solution. In addition, you should have an understanding of how DDD can play a role, no matter which style of architecture you choose to adopt.

In the next section, we will look to apply all we have learned in this and previous chapters to a real-world business use case. We will apply both the strategic and tactical patterns of DDD to break a complex domain into subdomains and bounded contexts and iteratively build a solution, using technologies that are based on the Java programming language.

Part 2: Real-World DDD

In the first section of this book, we looked at the vocabulary of **Domain-Driven Design (DDD)** and how it fits in the context of commonly used architecture styles and programming paradigms. In this section, we will implement a real-world application, starting from business motivations and requirements, and employ a bunch of techniques and practices that enable us to apply the tenets of DDD's strategic and tactical design.

This part contains the following chapters:

3
Understanding
the Domain

"A spoon does not know the taste of soup, nor a learned fool the taste of wisdom."

– Welsh proverb

In this chapter, we will introduce a fictitious organization named **Kosmo Prima** (**KP**) Bank that is looking to modernize its product offerings in the international trade business. In order to establish a business strategy that sets it up for sustained success in the medium to long term, we will employ a series of techniques and practices to help expedite its path from strategy to execution.

At the outset, let's gain a high-level understanding of KP Bank's business domain before we dive deeper.

In this chapter, we will cover the following topics:

- The domain of international trade
- International trade at KP Bank
- Understanding international trade strategy at KP Bank
- International trade products and services

At the end of this chapter, you will gain an appreciation of how to employ techniques such as the business value canvas and the lean canvas to establish a sound understanding of business strategy. Furthermore, we will examine how plotting an impact map will allow us to correlate business deliverables to goals. Finally, the Wardley mapping exercise will establish the importance of our business decisions in relation to our competitive landscape.

The domain of international trade

In many countries, international trade represents a significant portion of the **gross domestic product (GDP)**, making an exchange of capital, goods, and services between untrusted parties spread across the globe a necessity. While economic organizations such as the **World Trade Organization (WTO)** were formed specifically to ease and facilitate this process, differences in factors such as economic policy, trade laws, and currency ensure that carrying out trade internationally can be a complex process, with several entities involved across countries. A letter of credit exists to simplify this process. Let's take a look at how it works.

International trade at KP Bank

KP Bank has been in business for several years and has been focusing on providing a variety of banking solutions, such as retail, corporate, securities, and other products. They have been steadily expanding operations to other countries and continents. This has allowed them to expand their international trade business significantly in the last decade. While they have been among the leaders in this space, the recent onset of new digital-native competitors has started to eat into their business and impact their top line adversely. Customers are complaining that the process is too cumbersome, time-consuming, and lately unreliable. In addition, due to a very inefficient manual process that is currently in place, KP Bank has been finding it very hard to keep a check on costs. In just the last 3 years, they have had to increase transaction processing costs by around 50 percent! Not surprisingly, this has coincided with plummeting customer satisfaction, which is evidenced by the fact that the number of customers serviced has remained flat over the intervening time.

The CIO has recognized that there is a need to look at this problem afresh and come up with a strategy that sets KP Bank up for sustained success for the next few years and reestablish them as one of the leaders in the international trade business.

Understanding international trade strategy at KP Bank

To arrive at an optimal solution, it is important to have a strong appreciation of the company's business goals and their alignment to support the needs of the users of the solution. We will introduce a set of tools and techniques we have found to be useful.

> **Note**
>
> It is pertinent to note that these tools were conceived independently, but when practiced in conjunction with other DDD techniques, they can accentuate the effectiveness of the overall process and solution. The use of these tools should be considered complementary to your DDD journey.

Let's look at some of the most popular techniques we have employed to quickly gain an understanding of the business problem and propose solutions.

The business model canvas

As we have mentioned several times, it is important to make sure that we are solving the right problem before attempting to solve it. The business model canvas, originally conceived by Swiss consultant Alexander Osterwalder as part of his PhD thesis, is a quick and easy way to establish that we are solving a valuable problem in a single visual that captures nine elements of your business:

- **Value propositions**: What do you do?
- **Key activities**: How do you do it?
- **Key resources**: What do you need?
- **Key partners**: Who will help you?
- **Cost structure**: What will it cost?
- **Revenue streams**: How much will you make?
- **Customer segments**: Who are you creating value for?
- **Customer relationships**: Who do you interact with?
- **Channels**: How do you reach your customers?

The business model canvas helps establish a shared understanding of the big picture among a varied set of groups, including business stakeholders, domain experts, product owners, architects, and developers. We have found it very useful when embarking on both greenfield and brownfield engagements. Here is an attempt we made to create a business model canvas for the international trade business at KP Bank:

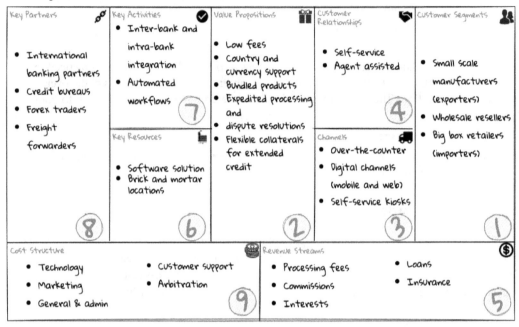

Figure 3.1 – A business model canvas

Using this canvas leads to insights about the customers we intend to serve at the bank, what value propositions are offered through what channels, and how we make money. When developing a business model canvas, it is recommended that we follow the numbered sequence depicted in the previous figure in order to gain a better understanding of the following:

- The desirability of the business (who our customers are and what they want)

- The feasibility of the business (how we can operationalize and deliver it)

- The economic viability of the business (how we can identify costs and capture profits)

Creating a business model canvas can prove challenging if you do not have an existing product already, which is usually true in the case of start-ups or existing enterprises expanding into new business areas. In such cases, a variation in the form of the lean canvas is worth exploring.

The lean canvas

A variation of the business model canvas called the **lean canvas** was conceived by Ash Maurya for lean startups. In contrast to the business model canvas, the main emphasis here is to first and foremost elaborate on the problem that needs to be solved and explore potential solutions. In order to make the canvas actionable, the idea is to capture items that were most uncertain and/or risky. This is pertinent for businesses operating under high uncertainty (which is usually true for start-ups). Similar to DDD, it encourages you to focus on an issue as the starting point for building a business.

Structurally, it is similar to the business model canvas, with the following differences:

- **Problems** instead of *key partners*: It is common for businesses to fail due to them misunderstanding the problem they are solving. The rationale for replacing the **key partners** block is that when you are an unknown entity looking to establish an unproven product, pursuing key partnerships may be premature.

- **Solutions** instead of *key activities*: It is important to try more than one solution iteratively and respond to feedback. **Key activities** are removed because they are usually a by-product of the solution.

- **Key metrics** instead of *key resources*: It is very important to know that we are progressing in the right direction. It is advisable to focus on a small number of **key metrics** to enable pivoting quickly if needed. **Key resources** have become relatively easy (with the advent of the cloud and the availability of mature frameworks). Furthermore, they may appear in the **Unfair advantage** box, which we will discuss next.

- **Unfair advantage** instead of *customer relationships*: This clearly establishes our differentiators that are hard to replicate. This is closely aligned to the idea of the core subdomain that we discussed in *Chapter 1, The Rationale for Domain-Driven Design*, and gives us a clear picture of what we need to focus our energies on at the outset.

The result of a lean canvas workshop we ran for KP Bank is shown here:

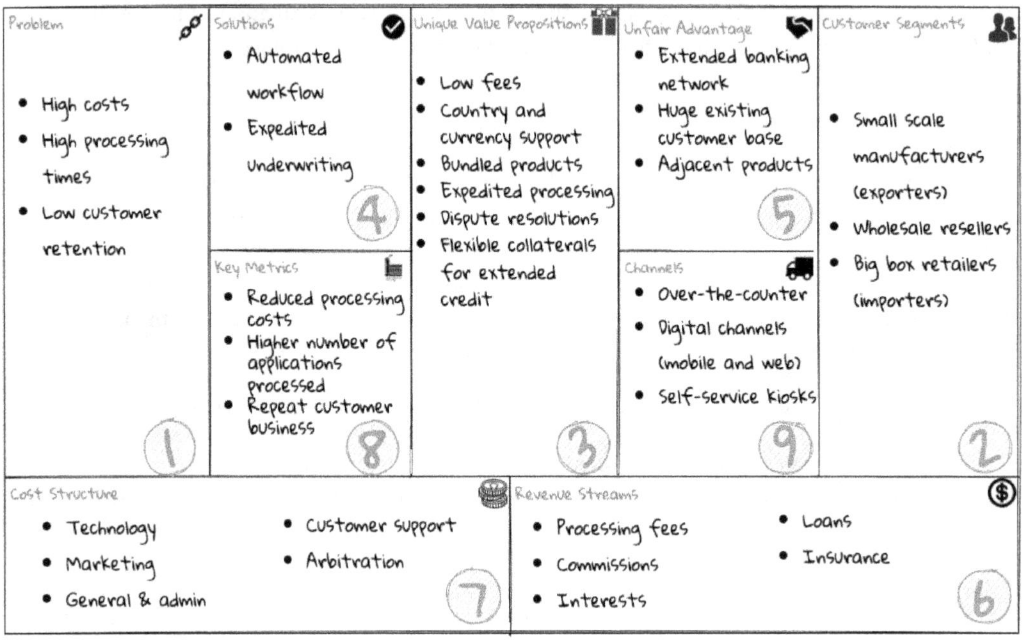

Problem	Solutions	Unique Value Propositions	Unfair Advantage	Customer Segments
• High costs • High processing times • Low customer retention	• Automated workflow • Expedited underwriting **4** **Key Metrics** • Reduced processing costs • Higher number of applications processed • Repeat customer business **8**	• Low fees • Country and currency support • Bundled products • Expedited processing • Dispute resolutions • Flexible collaterals for extended credit **3**	• Extended banking network • Huge existing customer base • Adjacent products **5** **Channels** • Over-the-counter • Digital channels (mobile and web) • Self-service kiosks **9**	• Small scale manufacturers (exporters) • Wholesale resellers • Big box retailers (importers) **2**
1				

Cost Structure		Revenue Streams	
• Technology • Marketing • General & admin	• Customer support • Arbitration **7**	• Processing fees • Commissions • Interests	• Loans • Insurance **6**

Figure 3.2 – A lean canvas for the international trade business

The exact sequence in which to fill out the lean canvas may vary. On his blog, Ash Maurya suggests that there may not be a prescriptive running order to do this exercise (`https://blog.leanstack.com/what-is-the-right-fill-order-for-a-lean-canvas/`). Personally, we like starting by elaborating on the problem before moving on to other aspects of the canvas. Both the business model canvas and the lean canvas provide a high-level view of the business model, the high-priority problems, and the potential solutions. Next, let's look at impact mapping, another lightweight planning technique to arrive at an outcome-driven plan, based on mind maps.

Impact maps

An impact map is a visualization and strategic planning tool that enables you to understand scope and underlying assumptions. It was created collaboratively by senior technical and business people in the form of a mind map by considering the following four aspects:

- **Goals**: *Why* are we doing this?

- **Actors**: *Who* are the consumers or users of our product? In other words, who will be impacted by it?

- **Impacts**: *How* can the consumers' change in behavior help us achieve our goals – in other words, the impacts that we're trying to create?

- **Deliverables**: *What* can we do, as an organization or a delivery team, to support the required impacts – in other words, the software features or process changes that are required to be realized as part of the solution?

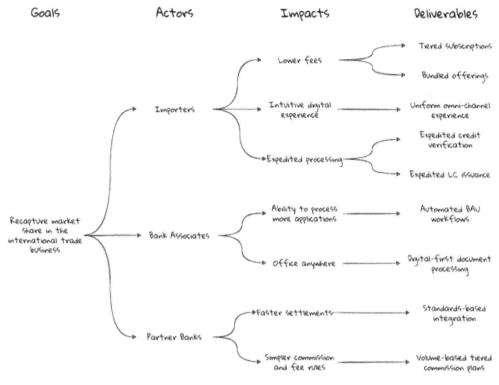

Figure 3.3 – A simple impact map

Impact mapping provides an easy-to-understand visual representation of the relationship between the goals, users, and impacts on the deliverables. Next, let us examine Wardley maps that enable us to dive deeper, understand our purpose, and determine which portions of the business provide the most value.

Wardley maps

The business model canvas and lean canvas can help establish clarity of purpose at a high level. The Wardley map is another tool to help build a business strategy and establish purpose. It provides a sketch of the people that a system is built for, followed by the benefits the system offers them, and a chain of needs required to provide those benefits (called the **value chain**). Next, the value chain is plotted along an evolution axis, which ranges from something that is uncharted and uncertain to something that is highly standardized. Building a Wardley map can be done in six steps:

1. **Purpose**: What is your purpose? Why does the organization or project exist?

2. **Scope**: What is (and not) included within the scope of the map?

3. **Users**: Who uses or interacts with the thing you are mapping?

4. **User needs**: What do your users need from the thing you are mapping?

5. **Value chain**: What do we need to be doing to fulfill those needs captured previously? These needs are arranged according to their dependencies, resulting in the creation of a value chain that maps user needs to a series of activities in the order of their visibility to the user (going from most visible to the least).

6. **Map**: Finally, plot the map using the evolutionary characteristics to decide where to place each component along the horizontal axis.

We conducted a Wardley mapping exercise at KP Bank for their international trade business, as shown here:

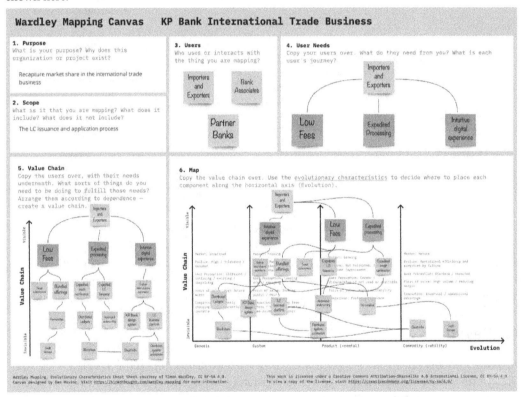

Figure 3.4 – A Wardley map for the international trade business at KP Bank

> **Note**
> On this canvas, we have chosen to elaborate on the needs of only one class of users (importers and exporters) for brevity. In a real-world scenario, we would have to repeat *steps 4, 5, and 6* for all types of users.

The Wardley map makes it easy to understand the capabilities provided by our solution, their dependencies, and how value is derived. It also helps depict how these capabilities play out compared to those offered by competitors, allowing you to prioritize attention appropriately and make build versus buy decisions.

We have examined several lightweight and collaborative techniques to quickly understand the problem space and the impacts we can have on our users and our business. Each of these techniques is fairly lightweight and can be completed within a matter of a few hours. Each enables us to focus on the most impactful business areas and maximize ROI. In our experience, it is worth experimenting with more than one of these exercises (potentially all of them) as each can highlight a different facet of the business/user needs.

International trade products and services

International trade is fraught with risk, which then presents a degree of uncertainty over the timing of payments between the seller (exporter) and the buyer (importer), especially due to a lack of trust between the parties involved. For exporters, until payment is received, all sales are gifts. Consequently, exporters prefer receiving payment as soon as the order is placed or at least before the goods are shipped. For importers, until the goods are received, all payments made toward a purchase are donations. Consequently, importers prefer receiving goods as soon as possible and delaying payment until the goods are resold to generate enough money to pay the seller.

This situation presents an opportunity for trusted intermediaries (such as KP Bank) to play a significant role in brokering international trade transactions in a secure manner. KP bank offers a number of products to facilitate international trade payments, as listed here:

- Letter of credit (LC)
- Documentary collections (DC)
- Open account
- Cash-in-advance
- Consignments

The following diagram shows the risk profile of each of these payment methods from both an exporter's and importer's perspective:

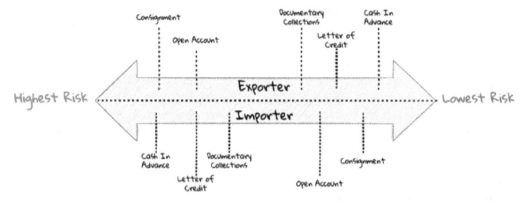

Figure 3.5 – The risk profile of international trade payment methods

As is evident here, the DC and LC products offer a good balance in providing solutions that are relatively secure from the perspective of both parties. The involvement of a trusted intermediary such as KP Bank is required to play a part in the fulfillment process makes these payment methods less risky for both parties. From the bank's perspective, streamlining the process for these products as a priority also provides a greater business opportunity compared to the other products. Of the two, the LC product satisfies most criteria, outlined against user needs in the recently concluded business strategy sessions we elaborated on earlier. Hence, the stakeholders at KP Bank have decided to move forward with heavy investment in the LC product at the outset.

In the next chapter, and indeed the rest of the book, we will elaborate on how we can improve the LC application, issuance, and related processes by making use of tenets that align closely with the tenets of DDD.

Summary

In this chapter, we have explored a variety of techniques that help establish whether a particular problem is the right one to be solved. Specifically, we looked at the business value canvas and the lean canvas to clarify the business strategy for both start-ups and established enterprises. We then looked at impact maps that enable you to unambiguously correlate business goals to user impacts and the deliverables needed to create those impacts. Finally, we looked at Wardley maps to further drill down areas that are important to focus energies on, including establishing build versus buy decisions, the importance of business strategy in relation to competitors, and the relative risk involved when heading into uncharted waters.

In the next chapter, we will look at techniques and practices to drill down further and gain an understanding of the LC business so that we can start crafting domain model(s) to enable us to arrive at an appropriate solution.

Further reading

Learn more about the lean canvas at `https://blog.leanstack.com/what-is-the-right-fill-order-for-a-lean-canvas/`.

4

Domain Analysis and Modeling

"He who asks a question remains a fool for 5 minutes. He who does not ask remains a fool forever."

– Chinese proverb

As we saw in the previous chapter, misinterpreted requirements can cause a significant portion of software projects to fail. Arriving at a shared understanding and creating a useful domain model requires high degrees of collaboration between domain experts. In this chapter, we will introduce the sample application we will use throughout the book and explore modeling techniques, such as domain storytelling and EventStorming, to enhance our collective understanding of a problem in a reliable and structured manner.

The following topics will be covered in this chapter:

- Introducing the example application (a letter of credit)
- Enhancing shared understanding
- Domain storytelling
- EventStorming

This chapter will help developers and architects learn how to apply these techniques in real-life situations to produce elegant software solutions that mirror the domain problem that needs to be solved. Similarly, non-technical domain experts will understand how to communicate their ideas and collaborate effectively with technical team members to accelerate the process of arriving at a shared understanding.

Technical requirements

There are no specific technical requirements for this chapter. However, given that it may become necessary to collaborate remotely as opposed to being in the same room with access to a whiteboard, it will be useful to have access to the following:

- A digital whiteboard (such as `https://www.mural.co/` or `http://miro.com/`)
- An online domain storytelling modeler (such as `https://www.wps.de/modeler/`)

Understanding a letter of credit

A documentary **Letter of Credit (LC)** is a financial instrument issued by banks as a contract between an importer (or buyer) and an exporter (or seller). This contract specifies the terms and conditions of the transaction, under which the importer promises to pay the exporter in exchange for the goods or services provided by the exporter. An LC transaction typically involves multiple parties. A simplified summary of the parties involved is described as follows:

- **Importer**: The buyer of the goods or services.
- **Exporter**: The seller of the goods or services.
- **Freight forwarder**: The agency that handles the shipment of goods on behalf of the exporter. This is only applicable in cases where there is an exchange of physical goods.
- **Issuing bank**: The bank that the importer requests to issue the LC application. Usually, the importer has a preexisting relationship with this bank.
- **Advising bank**: The bank that informs the exporter about the issuance of the LC. This is usually a bank that is native to the exporter's country.
- **Negotiating bank**: The bank that the exporter submits documents for the shipment of goods, or the services provided. Usually, the exporter has a preexisting relationship with this bank.
- **Reimbursement bank**: The bank that reimburses the funds to the negotiating bank, at the request of the issuing bank.

> **Note**
> One bank can play more than one role in a given transaction. In the most complex cases, there can be four distinct banks involved in a transaction (sometimes even more, but we will skip those cases for brevity).

An LC issuance application

As discovered in the previous chapter, Kosmo Primo Bank needs us to focus on streamlining the process used for LC application and issuance functions. In this chapter, and indeed the rest of this book, we will strive to understand, evolve, design, and build a software solution to make the process more efficient by replacing the largely manual and error-prone workflows with more simplified processes, based on larger amounts of automation.

We understand that unless you are an expert dealing with international trade, it is unlikely that you will have an intimate understanding of concepts such as LCs. In the upcoming section, we will look at demystifying LCs and how to work with them.

Enhancing shared understanding

When working with a problem where domain concepts are unclear, there is a need to arrive at a common understanding among key team members (both those that have bright ideas – the business/ product people, and those that translate those ideas into working software – the software developers). For this process to be effective, we tend to look for approaches that are as follows:

- Quick, informal, and effective
- Collaborative – easy to learn and adopt for both non-technical and technical team members
- Pictorial, because a picture can be worth a thousand words
- Usable for both coarse-grained and fine-grained scenarios

There are several means to arrive at this shared understanding. The following are some of the commonly used approaches:

- UML
- BPMN
- Use cases
- User story mapping
- CRC models
- Data flow diagrams

These modeling techniques try to formalize knowledge and express it in the form of a diagram or text to help deliver business requirements as a software product. However, this attempt has not narrowed but widened the gap between business and software systems. While these methods tend to work well for technical audiences, they are usually not as appealing to non-technical users.

In order to restore the balance and promote the use of techniques that can work for both parties, we will use **domain storytelling** and **EventStorming** as our means to capture business knowledge from domain experts for the consumption of developers and business analysts, among others.

Domain storytelling

Scientific research has now proven that learning methods that employ audiovisual aids assist both teachers and students in retaining and internalizing concepts very effectively. In addition, teaching what we have learned helps reinforce ideas and stimulate the formation of new ones.

Domain storytelling is a collaborative modeling technique that combines a pictorial language, real-world examples, and a workshop format to serve as a very simple, quick, and effective technique for sharing knowledge among team members. Domain storytelling is a technique invented and popularized by Stefan Hofer and Henning Schwentner, based on some related work done at the University of Hamburg called *cooperation pictures*.

A pictorial notation of the technique is illustrated in the following diagram:

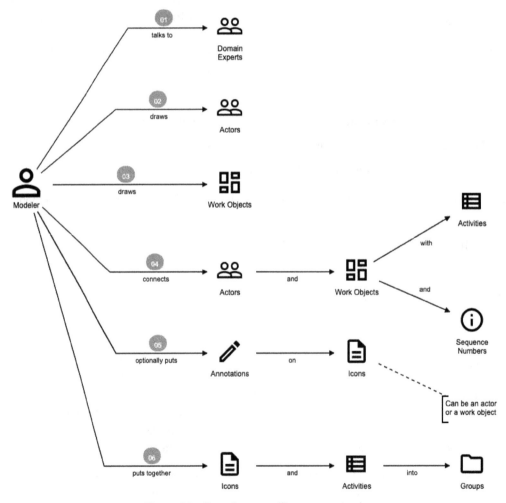

Figure 4.1 – Domain storytelling summarized

A domain story is conveyed using the following attributes:

- **Actors**: Stories are communicated from the perspective of an actor (noun) – for example, the issuing bank, which plays an active role in the context of a particular story. It is a good practice to use the ubiquitous language for the particular domain.

- **Work objects**: Actors act on some object – for example, applying for an LC. Again, this would be a term (noun) commonly used in the domain.

- **Activities**: Actions (verb) performed by the actor on a work object. Represented by a labeled arrow connecting the actor and the work object.

- **Annotations**: Used to capture additional information as part of the story, usually represented in a few sentences.

- **Sequence numbers**: Usually, stories are told one sentence after the other. Sequence numbers help capture the sequence of activities in a story.

- **Groups**: An outline to represent a collection of related concepts, ranging from repeated/optional activities to subdomains/organizational boundaries.

Using DST for an LC application

KP Bank has a process that allows the processing of LCs. However, this process is very archaic, paper-based, and manually intensive. Very few at the bank fully understand the process end to end, and natural attrition has meant that the process is overly complex without good reason. So, they are looking to digitize and simplify this process. DST itself is just a graphical notation that can be done in isolation. However, it is typical to not do this on your own and employ a workshop style, with domain experts and software experts working collaboratively.

In this section, we will employ a DST workshop to capture the current business flow. The following is an excerpt of such a conversation between *Katie, the domain expert*, and *Patrick, the software developer*:

Patrick: *Can you give me a high-level overview of a typical LC flow?*

Katie: *Sure, it all begins with the importer and the exporter entering into a contract for the purchase of goods or services.*

Patrick: *What form does this contract take? Is it a formal document clause? Or is this just a conversation?*

Katie: *This is just a conversation.*

Patrick: *Oh okay. What does the conversation cover?*

Katie: *Several things – the nature and quantity of goods, pricing details, payment terms, shipment costs and timelines, insurance, and warranty, among other things. These details can be captured in a purchase order – which is a simple document clause elaborating the aforementioned.*

At this time, Patrick draws this part of the interaction between the importer and the exporter. This graphic is depicted in the following diagram:

Figure 4.2 – The interaction between the importer and the exporter

Patrick: *This seems straightforward, so where does the bank come into the picture?*

Katie: *This is an international trade, and both the importer and the exporter need to mitigate the financial risk involved in such business transactions. So, they involve a bank as a trusted mediator.*

Patric: *What kind of bank is this?*

Katie: *Usually, there are multiple banks involved. But it all starts with an* **issuing bank.**

Patrick: *What is an issuing bank?*

Katie: *Any bank that is authorized to mediate international trade deals. This has to be a bank in the importer's country.*

Patrick: *Does the importer need to have an existing relationship with this bank?*

Katie: *Not necessarily. There may be other banks with whom the importer may have a relationship – which, in turn, liaise with the issuing bank on the importer's behalf. But to keep it simple, let's assume that the importer has an existing relationship with the issuing bank – which is our bank, in this case.*

Patrick: *Does the importer provide details of the purchase order to the issuing bank to get started?*

Katie: *Yes. The importer provides the details of the transaction by making an* **LC application.**

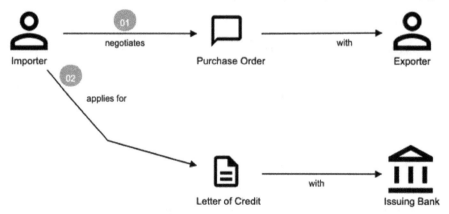

Figure 4.3 – Introducing the LC and the issuing bank

Patrick: *What does the issuing bank do when they receive this LC application?*

Katie: *Mainly two things – vet the financial standing of the importer and the legality of the goods being imported.*

Patrick: *Okay. What happens if everything checks out?*

Katie: *The issuing bank approves the LC and notifies the importer.*

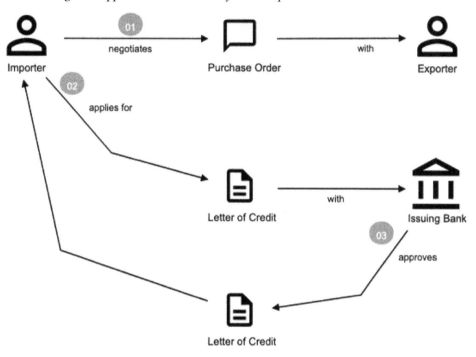

Figure 4.4 – Notifying LC approval to the importer

Patrick: *What happens next? Does the issuing bank contact the exporter now?*

Katie: *Not yet. It is not that simple. The issuing bank can only deal with a counterpart bank in the exporter's country. This bank is called the* **advising bank.**

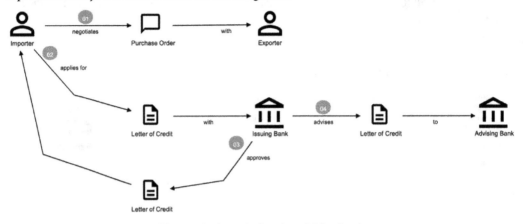

Figure 4.5 – Introducing the advising bank

Patrick: *What does the advising bank do?*

Katie: *The advising bank notifies the exporter about the LC.*

Patrick: *Doesn't the importer need to know that the LC has been advised?*

Katie: *Yes. The issuing bank notifies the importer that the LC has been advised to the exporter.*

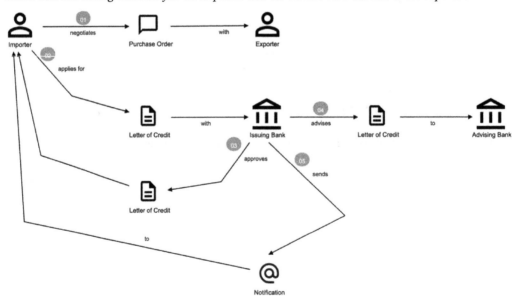

Figure 4.6 – Advice notification to the importer

Patrick: *How does the exporter know how to proceed?*

Katie: *Through the advising bank – they notify the exporter that the LC was issued.*

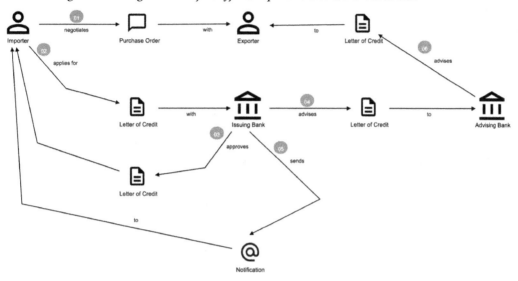

Figure 4.7 – Dispatching the advice to the exporter

Patrick: *Does the exporter initiate shipping at this time and how do they get paid?*

Katie: *Through the advising bank – they notify the exporter that the LC was issued, and this triggers the next steps in the process – this process of settling payment is called* **settlement**. *But let's focus on issuance right now. We will discuss settlement at a later time.*

We have now looked at an excerpt of a typical DST workshop. It has provided a reasonably good understanding of the high-level business flow. Note that we have not referenced any technical artifacts during the process.

To refine this flow and convert it into a form that can be used to design a software solution, we will need to further enhance this view. In the upcoming section, we will use **EventStorming** as a structured approach to achieve that.

EventStorming

> *"The amount of energy necessary to refute bullshit is an order of magnitude bigger than to produce it."*
>
> *– Alberto Brandolini*

Introducing EventStorming

In the previous section, we gained a high-level understanding of the LC issuance process. To be able to build a real-world application, it helps to use a method that delves into the next level of detail. EventStorming, originally conceived by Alberto Brandolini, is one such method for the collaborative exploration of complex domains.

In this method, you simply start by listing all the events that are significant to a business domain in roughly chronological order on a wall or whiteboard, using a bunch of colored sticky notes. Each of the note types (denoted by different colors) serves a specific purpose, as outlined here:

- **Domain event**: An event that is significant to the business process – expressed in the past tense.
- **Command**: An action or an activity that may result in one or more domain events occurring. This is either user-initiated or system-initiated, in response to a domain event.
- **User**: A person who performs a business action/activity.
- **Policy**: A set of business invariants (rules) that need to be adhered to for an action/activity to be successfully performed.
- **Query/read model**: A piece of information required to perform an action/activity.
- **External system**: A system significant to the business process but out of scope in the current context.
- **Hotspot**: A point of contention within the system that is likely confusing and/or puzzling beyond a small subsection of the team.
- **Aggregate**: An object graph whose state changes consistently and atomically. This is consistent with the definition of *aggregates* we saw in *Chapter 2, Where and How Does DDD Fit?*.

The depiction of the stickies for our EventStorming workshop is shown here:

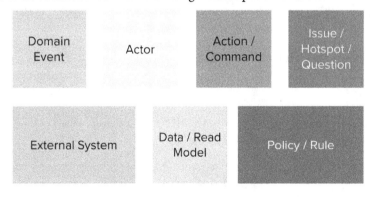

Figure 4.8 – An EventStorming legend

> **Why Domain Events?**
>
> When trying to understand a business process, it is convenient to explain significant facts or things in that context. This practice can also be informal and easy for uninitiated audiences. This provides an easy-to-digest visual representation of domain complexity.

Using EventStorming for the LC issuance application

Now that we have a high-level understanding of the current business process, thanks to the domain storytelling workshop, let's look at how we can delve deeper using EventStorming. The following is an excerpt of the stages from an EventStorming workshop for the same application:

1. **Outline the event chronology**: During this exercise, we recall significant **domain events** (using orange stickies) in the system and paste them on the whiteboard, as depicted in the following diagram. We ensure that the event stickies are pasted roughly in the chronological order of occurrence. As the timeline is enforced, the business flow will begin to emerge:

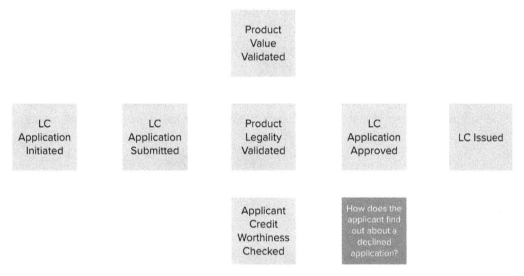

Figure 4.9 – Event chronology

This acts as an aid in understanding the big picture. This also enables people in the room to identify hotspots in the existing business process. In the preceding diagram, the process to handle *declined LC applications* is suboptimal – that is, the applicants do not receive any information when their application is declined.

To address this, we added a new domain event that explicitly indicates that an application is declined, as depicted in the following diagram:

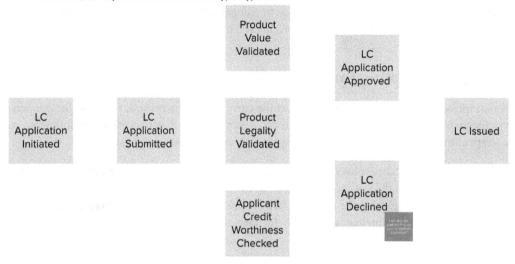

Figure 4.10 – A new event to handle declined applications

2. **Identify triggering activities and external systems**: Having arrived at a high-level understanding of event chronology, the next step is to embellish the visual with *activities/actions* that cause these events to occur (using blue stickies) and interactions with *external systems* (using pink stickies):

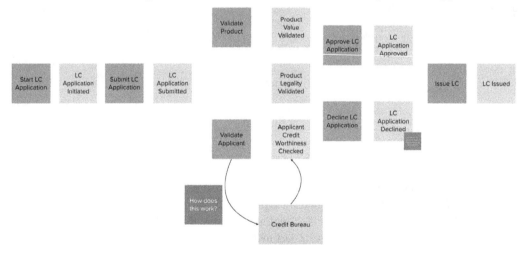

Figure 4.11 – Activities and external systems

3. **Capture users, context, and policies:** The next step is to capture *users* who perform these activities along with their functional *context* (using yellow stickies) and *policies* (using purple stickies).

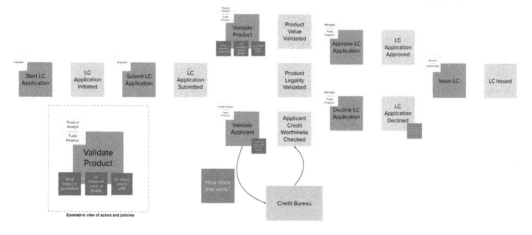

Figure 4.12 – Users and policies

4. **Outline query models:** Every activity requires a certain set of data. Users need to view out-of-band data that they need to act upon and also see the result of their actions. These sets of data are represented as *query models* (using green stickies):

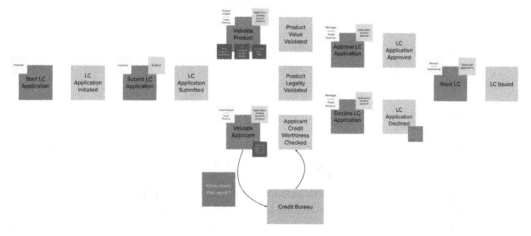

Figure 4.13 – A big picture EventStorming workshop board

Important note

For both the domain storytelling and EventStorming workshops, it works best when we have approximately six to eight people participating with the right mix of domain and technology experts.

This concludes the EventStorming workshop to gain a reasonably detailed understanding of the LC application and issuance process. Does this mean that we have concluded the domain requirements gathering process? Not at all – while we have made significant strides in understanding the domain, there is still a long way to go. The process of elaborating domain requirements is perpetual. Where are we in this continuum? The following diagram is an attempt to clarify:

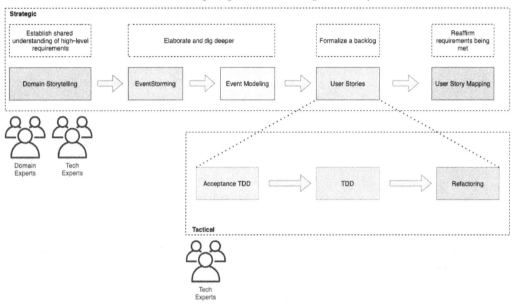

Figure 4.14 – Elaborating the domain requirements continuum

In subsequent chapters, we will examine the other techniques in more detail.

Summary

In this chapter, we examined two ways to enhance our collective understanding of a problem domain using two lightweight modeling techniques – domain storytelling and EventStorming.

Domain storytelling uses a simple pictorial notation to share business knowledge among domain experts and technical team members. EventStorming, on the other hand, uses a chronological ordering of domain events that occur as part of the business process to gain that same shared understanding.

Domain storytelling can be used as an introductory technique to establish a high-level understanding of a problem space, while EventStorming can be used to inform detailed design decisions of a solution space.

With this knowledge, we should be able to dive deeper into the technical aspects of solution implementation. In the next chapter, we will start the implementation of business logic and model our aggregate, along with commands and domain events.

Further reading

Title	Author	Location
Domain Storytelling	Stefan Hofer and Henning Schwentner	`https://leanpub.com/domainstorytelling`
An Introduction to Domain Storytelling	Virtual Domain-Driven Design	`https://www.youtube.com/watch?v=d9k9Szkdprk`
Domain storytelling resources	Stefan Hofer	`https://github.com/hofstef/awesome-domain-storytelling`
Introducing EventStorming	Alberto Brandolini	`https://leanpub.com/introducing_EventStorming`
Introducing Event Storming	Alberto Brandolini	`https://ziobrando.blogspot.com/2013/11/introducing-event-storming.html`
Event Storming for fun and profit	Dan Terhorst-North	`https://speakerdeck.com/tastapod/event-storming-for-fun-and-profit`
Event Storming Shopping List	Allen Holub	`https://holub.com/event-storming/`

5
Implementing Domain Logic

To communicate effectively, the code must be based on the same language used to write the requirements – the same language that the developers speak with each other and with domain experts.

– Eric Evans

In the *Command Query Responsibility Segregation (CQRS)* section of the book, we described how DDD and CQRS complement each other and how the command side (write requests) is the home of business logic. In this chapter, we will implement the command-side API for the **Letter of Credit (LC)** application using Spring Boot, Axon Framework, JSR-303 bean validations, and persistence options by contrasting between state-stored and event-sourced aggregates. The list of topics to be covered is as follows:

- Identifying aggregates
- Handling commands and emitting events
- Test-driving the application
- Persisting aggregates
- Performing validations

By the end of this chapter, you will have learned how to implement the core of your system (the domain logic) in a robust, well-encapsulated manner. You will also learn how to decouple your domain model from persistence concerns. Finally, you will be able to appreciate how to perform DDD's tactical design, using services, repositories, aggregates, entities, and value objects.

Technical requirements

To follow the examples in this chapter, you will need access to the following:

- JDK 1.8+ (we have used Java 16 to compile sample sources)
- Maven 3.x
- Spring Boot 2.4.x
- JUnit 5.7.x (included with Spring Boot)
- Axon Framework 4.4.7 (DDD and CQRS framework)
- Project Lombok (to reduce verbosity)
- Moneta 1.4.x (money and currency reference implementation – JSR 354)

Continuing our design journey

In the previous chapter, we discussed eventstorming as a lightweight method to clarify business flows. As a reminder, this is the output produced from our eventstorming session:

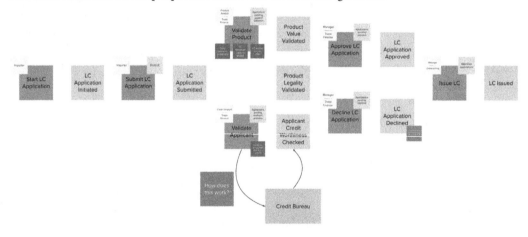

Figure 5.1 – Recap of the eventstorming session

As mentioned previously, the *blue* stickies in this diagram represent *commands*. We will be using the **Command Query Responsibility Segregation (CQRS)** pattern as a high-level architecture approach to implement the domain logic for our LC issuance application. Let's examine the mechanics of using CQRS and how it can result in an elegant solution. For a recap of what CQRS is and when it is appropriate to apply this pattern, please refer to the *When to use CQRS* section in *Chapter 2, Where and How Does DDD Fit?*.

> **Important Note**
>
> CQRS is by no means a silver bullet. Although it is general-purpose enough to be used in a variety of scenarios, it is a paradigm shift as applied to mainstream software problems. Like any other architecture decision, you should apply due diligence when choosing to adopt CQRS to your situation.

Let's look at how this works in practice by implementing a representative sliver of the command side of the LC application using the Spring and Axon frameworks.

Implementing the command side

In this section, we will focus on implementing the command side of the application. This is where we expect all the business logic of the application to be implemented. Logically, it looks like the following figure:

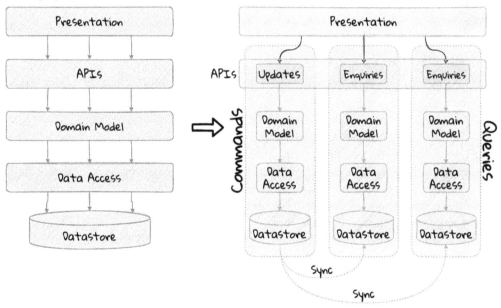

Figure 5.2 – Traditional versus CQRS architecture

The high-level sequence on the command side is described here:

1. A request to mutate state (command) is received.

2. In an event-sourced system, the command model is constructed by replaying existing events that have occurred for that instance. In a state-stored system, we would simply restore state by reading state from the persistence store.

3. If business invariants (validations) are satisfied, one or more domain events are readied with the intention to be published.

4. In an event-sourced system, the domain event is persisted on the command side. In a state-stored system, we would update the state of the instance in the persistence store.

5. The external world is notified by publishing these domain events onto an event bus. The event bus is an infrastructure component to which events are published.

Let's look at how we can implement this in the context of our LC issuance application.

> **Important Note**
>
> We depict multiple read models because it is possible (but not necessary) to create more than one read model, depending on the kinds of query use cases that need to be supported.

For this to work predictably, the read model(s) need to be kept in sync with the write models (we will examine some of the techniques to do that in detail later).

Tooling choices

Implementing CQRS does not require the use of any framework. Greg Young, who is considered the father of the CQRS pattern, advises against rolling our own CQRS framework in the essay found at `https://ordina-jworks.github.io/domain-driven%20design/2016/02/02/A-Decade-Of-DDD-CQRS-And-Event-Sourcing.html`, which is worth taking a look at. Using a good framework can help enhance developer effectiveness and accelerate the delivery of business functionality, while abstracting the low-level plumbing and non-functional requirements without limiting flexibility. In this book, we will make use of Axon Framework (`http://axonframework.org/`) to implement application functionality, as we have real-world experience in using it in large-scale enterprise development. There are other frameworks that work comparably, such as the Lagom framework (`https://www.lagomframework.com/`) and Eventuate (`https://eventuate.io/`), which are worth exploring as well.

Bootstrapping the application

To get started, let's create a simple Spring Boot application. There are several ways to do this. You can always use the Spring starter application at `https://start.spring.io` to create this application. Here, we will make use of the Spring CLI to bootstrap the application.

> **Important Note**
>
> To install the Spring CLI for your platform, please refer to the detailed instructions at `https://docs.spring.io/spring-boot/docs/current/reference/html/getting-started.html#getting-started.installing`.

To bootstrap the application, use the following command:

```
1  spring init \
2        --dependencies 'web,data-jpa,lombok,validation,h2,actuator' \
3        --name lc-issuance-api \
4        --artifactId lc-issuance-api \
5        --groupId com.example.api \
6        --packaging jar \
7        --description 'LC Issuance API' \
8        --package-name com.example.api \
9        --force
```

This should create a file named `lc-issuance-api.zip` in the current directory. Unzip this file to a location of your choice and add a dependency on Axon Framework in the `dependencies` section of the `pom.xml` file:

```
1        <dependency>
2            <groupId>org.axonframework</groupId>
3            <artifactId>axon-spring-boot-starter</artifactId>
4            <version>${axon-framework.version}</version>  ❶
5        </dependency>
```

1. You may need to change the version. We are at version 4.5.3 at the time of writing this book.

Also, add the following dependency on the `axon-test` library to enable unit testing of aggregates:

```
1        <dependency>
2            <groupId>org.axonframework</groupId>
3            <artifactId>axon-test</artifactId>
4            <scope>test</scope>
5            <version>${axon-framework.version}</version>
6        </dependency>
```

With the preceding setup, you should be able to run the application and start implementing the LC issuance functionality.

Let's look at how to implement these commands using Axon Framework.

Identifying commands

From the eventstorming session in the previous chapter, we have the following commands to start with:

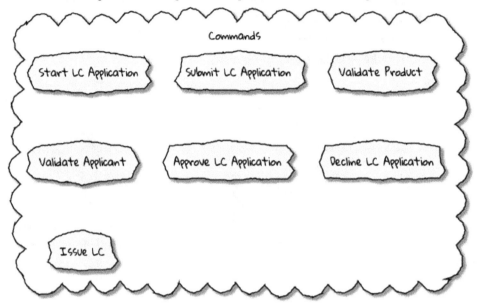

Figure 5.3 – Identified commands

Commands are always directed to an aggregate (the root entity) for processing (handling). This means that we need to resolve each of these commands to be handled by an aggregate. While the sender of the command does not care which component within the system handles it, we need to decide which aggregate will handle each command. It is also important to note that any given command can only be handled by a single aggregate within the system. Let's look at how to group these commands and assign them to aggregates. To be able to do that, we need to identify the aggregates in the system first.

Identifying aggregates

Looking at the output of the eventstorming session of our LC application, one potential grouping can be as follows:

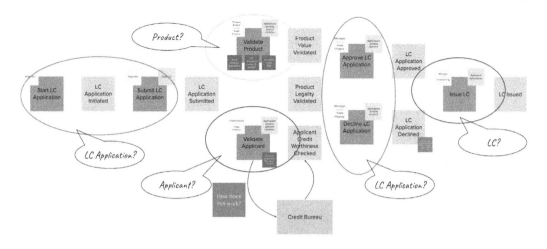

Figure 5.4 – First cut attempt at aggregate design

Some or all of these entities may be aggregates (For a more detailed explanation on the difference between aggregates and entities , please refer to *Chapter 1, The Rationale for Domain-Driven Design*). At first glance, it appears that we have four potential entities to handle these commands:

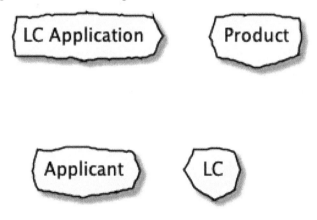

Figure 5.5 – Potential aggregates at first glance

At first glance, each of these entities may be classified as aggregates in our solution. Here, LC Application feels like a reasonably good choice for an aggregate, given that we are building a solution to manage LC applications. However, do the others make sense to be classified as such? Product and Applicant look like potential entities, but we need to ask ourselves whether we will need to operate on these outside of the purview of LC Application. If the answer is *yes*, then Product and Applicant *may* be classified as aggregates. But both Product and Applicant do not seem to require being operated on without an enveloping LC Application aggregate within this bounded context. It feels that way because both product and applicant details are required to be provided as part of the LC application process. At least from what we know of the process thus far, this seems to be true. This means that we are left with two potential aggregates – LC and LC Application:

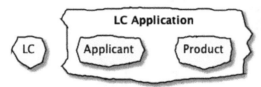

Figure 5.6 – Relationship between bounded contexts

When we look at the output of our eventstorming session, the LC Application aggregate transitions to become an LC aggregate much later in the life cycle. Let's work on the LC Application right now and suspend further analysis on the need for an LC aggregate to a later time.

> **Important Note**
>
> Colloquially, the terms aggregate and aggregate root are sometimes used interchangeably to mean the same thing. Aggregates can be hierarchical, and it is possible for aggregates to contain child aggregates. While both aggregates and aggregate roots handle commands, only one aggregate can exist as the root in a given context, and it encapsulates access to its child aggregates, entities and value objects.

It is important to note that entities may be required to be treated as aggregates in a different bounded context and this kind of treatment is entirely context dependent.

When we look at the output of our eventstorming session, the LC Application transitions to become an LC much later in the lifecycle in the Issuance context. Our focus right now is to optimize and automate the LC application flow of the overall issuance process. Now that we have settled on working with the LC Application aggregate (root), let's start writing our first command to see how this manifests itself in code.

Test-driving the system

While we have a reasonably good conceptual understanding of the system, we are still in the process of refining this understanding. Test-driving the system allows us to exercise our understanding by acting as the first client of the solution that we are producing.

> **Important Note**
>
> The practice of test-driving the system is very well illustrated in the best-selling book *Growing Object-Oriented Software, Guided by Tests* by authors Nat Price and Steve Freeman. This is worth looking at to gain a deeper understanding of this practice.

So, let's start with the first test. To the external world, an event-driven system typically works in a manner depicted in the following figure:

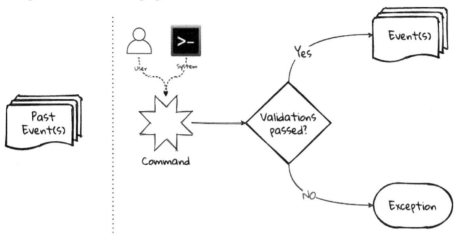

Figure 5.7 – An event-driven system

This figure can be explained as follows:

1. An optional set of domain events may have occurred in the past.

2. A command is received by the system (initiated manually by a user or automatically by a part of the system), which acts as a stimulus.

3. The command is handled by an aggregate, which then proceeds to validate the received command to enforce invariants (structural and domain validations).

4. The system then reacts in one of two ways:

    A. Emit one or more events.

    B. Throw an exception.

> **Important Note**
>
> The code snippets shown in this chapter are excerpts to highlight significant concepts and techniques. For the full working example, please refer to the accompanying source code for this chapter (included in the ch05 directory).

Axon Framework allows us to express tests in the following form:

```
 1  public class LCApplicationAggregateTests {
 2
 3      private FixtureConfiguration<LCApplication> fixture;                    ❶
 4
 5      @BeforeEach
 6      void setUp() {
 7          fixture = new AggregateTestFixture<>(LCApplication.class);          ❷
 8      }
 9
10      @Test
11      void shouldPublishLCApplicationCreated() {
12          fixture.given()                                                     ❸
13
14                  .when(new CreateLCApplicationCommand())                     ❹
15
16                  .expectEventsMatching(exactSequenceOf(                      ❺
17                          messageWithPayload(any(LCApplicationCreatedEvent.class)),  ❻
18                          andNoMore()                                         ❼
19                  ));
20      }
21  }
```

1. `FixtureConfiguration` is an Axon Framework utility to aid the testing of aggregate behavior using a BDD-style given-when-then syntax.

2. `AggregateTestFixture` is a concrete implementation of `FixtureConfiguration` where you need to register your aggregate class – in our case, `LCApplication` is the candidate to handle commands directed to our solution.

3. Since this is the start of the business process, there are no events that have occurred thus far. This is signified by the fact that we do not pass any arguments to the given method. In other examples we will discuss later, there will likely be events that have already occurred prior to receiving this command.

4. This is where we instantiate a new instance of the command object. Command objects are usually similar to data transfer objects, carrying a set of information. This command will be routed to our aggregate for handling. We will take a look at how this works in detail shortly.

5. Here, we are declaring that we expect events matching an exact sequence.

6. Here, we are expecting an event of the `LCApplicationCreated` type to be emitted as a result of successfully handling the command.

7. We are finally saying that we do not expect any more events, which means that we expect exactly one event to be emitted.

Implementing the command

`CreateLCApplicationCommand` in the previous simplistic example does not carry any state. Realistically, the command will likely look something like what is depicted as follows:

```
1    import lombok.Data;
2
3    @Data
4    public class CreateLCApplicationCommand {    ❶
5
6        private LCApplicationId id;                ❷
7        private ClientId clientId;
8        private Party applicant;                   ❸
9        private Party beneficiary;
10       private AdvisingBank advisingBank;         ❸
11       private LocalDate issueDate;
12       private MonetaryAmount amount;
13       private String merchandiseDescription;
14
15   }
```

1. This is the command class. When naming commands, we typically use an imperative style; that is, they usually begin with a verb denoting the action required. Note that this is a data transfer object. In other words, it is simply a bag of data attributes. Also note how it is devoid of any logic (at least at the moment).

2. This is the identifier for the LC application. We are assuming client-generated identifiers in this case. The topic of using server-generated versus client-generated identifiers is out of scope for the subject of this book. You may use either, depending on what is advantageous in your context. Also note that we are using a strong type for the LCApplicationId identifier as opposed to a primitive type, such as a numeric or a string value. It is also common in some cases to use UUIDs as an identifier. However, we prefer using strong types to be able to differentiate between identifier types. Note how we are using a ClientId type to represent the creator of the application.

3. The Party and AdvisingBank types are complex types to represent those concepts in our solution. Care should be taken to consistently use names that are relevant in the problem (business) domain as opposed to using names that only make sense in the solution (technology) domain. Note the attempt to make use of the ubiquitous language of the domain experts in both cases. This is a practice that we should always be conscious of when naming things in the system.

It is worth noting that merchandiseDescription is left as a primitive String type. This may feel contradictory to the commentary we presented previously. We will address this in the upcoming *Structural validations* section.

Now, let's see what the event we will emit as a result of successfully processing the command will look like.

Implementing the event

In an event-driven system, mutating system state by successfully processing a command usually results in a domain event being emitted to signal the state mutation to the rest of the system. A simplified representation of a real-world `LCApplicationCreatedEvent` event is shown here:

```
1   import lombok.Data;
2
3   @Data
4   public class LCApplicationCreatedEvent {    ❶
5
6       private LCApplicationId id;
7       private ClientId clientId;
8       private Party applicant;
9       private Party beneficiary;
10      private AdvisingBank advisingBank;
11      private LocalDate issueDate;
12      private MonetaryAmount amount;
13      private String merchandiseDescription;
14
15  }
```

1. When naming events, we typically use names in the past tense to denote things that have already occurred and are to be accepted unconditionally as empirical facts that cannot be changed.

You will likely notice that the structure of the event is currently identical to that of the command. While this is true in this case, it may not always be that way. The amount of information that we choose to disclose in an event is context-dependent. It is important to consult with domain experts when publishing information as part of events. You may choose to withhold certain information in the event payload. For example, consider `ChangePasswordCommand`, which contains the newly changed password. It might be prudent to not include the changed password in the resulting `PasswordChangedEvent`.

We have looked at the command and the resulting event in the previous test. Let's look at how this is implemented under the hood by looking at the aggregate implementation.

Designing the aggregate

The aggregate is the place where commands are handled and events are emitted. The good thing about the test that we have written is that it is expressed in a manner that hides the implementation details. But let's look at the implementation to be able to appreciate how we can get our tests to pass and meet the business requirement:

```
 1   public class LCApplication {
 2
 3       @AggregateIdentifier                                                    ❶
 4       private LCApplicationId id;
 5
 6       @SuppressWarnings("unused")
 7       private LCApplication() {
 8           // Required by the framework
 9       }
10
11       @CommandHandler                                                         ❷
12       public LCApplication(CreateLCApplicationCommand command) {              ❸
13           // TODO: perform validations here
14           AggregateLifecycle.apply(new LCApplicationCreatedEvent(command.getId())); ❹
15       }
16
17       @EventSourcingHandler                                                   ❺
18       private void on(LCApplicationCreatedEvent event) {
19           this.id = event.getId();
20       }
21   }
```

1. This is the aggregate identifier for the LCApplication aggregate. For an aggregate, the identifier uniquely identifies one instance from another. For this reason, all aggregates are required to declare an identifier and mark it to use the @AggregateIdentifier annotation provided by the framework.

2. The method that is handling the command needs to be annotated with the @CommandHandler annotation. In this case, the command handler happens to be the constructor of the class, given that this is the first command that can be received by this aggregate. We will see examples of subsequent commands being handled by other methods later in the chapter.

3. The @CommandHandler annotation marks a method as being a command handler. The exact command that this method can handle needs to be passed as a parameter to the method. Note that there can only be one command handler in the entire system for any given command.

4. Here, we are emitting LCApplicationCreatedEvent using the AggregateLifecycle utility provided by the framework. In this very simple case, we are emitting an event unconditionally on receipt of the command. In a real-world scenario, it is conceivable that a set of validations will be performed before deciding to either emit one or more events, or failing the command with an exception. We will look at more realistic examples later in the chapter.

5.   The need for `@EventSourcingHandler` and its role is likely very unclear at this time. We will explain the need for this in detail in an upcoming section of this chapter.

This was a whirlwind introduction to a simple event-driven system. We still need to understand the role of `@EventSourcingHandler`. To understand that, we will need to appreciate how aggregate persistence works and the implications it has on our overall design.

Persisting aggregates

When working with any system of even moderate complexity, we are required to make interactions durable; that is, interactions need to outlast system restarts, crashes, and so on. So the need for persistence is a given. While we should always endeavor to abstract persistence concerns from the rest of the system, our persistence technology choices can have a significant impact on the way we architect our overall solution. We have a couple of choices in terms of how we choose to persist aggregate state that are worth mentioning:

- State-stored
- Event-sourced

Let's examine each of these techniques in more detail in the following sections.

State-stored aggregates

Saving current values of entities is by far the most popular way to persist state – thanks to the immense popularity of relational databases and **Object-Relational Mapping (ORM)** tools such as Hibernate. And there is good reason for this ubiquity. Until recently, a majority of enterprise systems used relational databases almost as a default to create business solutions, with ORMs arguably providing a very convenient mechanism to interact with relational databases and their object representations. For example, for our `LCApplication`, it is conceivable that we could use a relational database with a structure that would look something like the following:

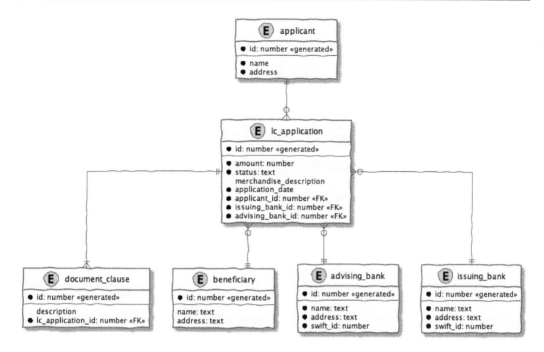

Figure 5.8 – Typical entity relationship model

Irrespective of whether we choose to use a relational database or a more modern NoSQL store – for instance, a document store, key-value store, column family store, and so on – the style we use to persist information remains more or less the same, which is to store the current values of the attributes of the said aggregate/entity. When the values of attributes change, we simply overwrite old values with newer ones; that is, we store the current state of aggregates and entities – hence the name *state-stored*. This technique has served us very well over the years, but there is at least one more mechanism that we can use to persist information. We will look at this in more detail next.

Event-sourced aggregates

Developers have also been relying on logs for a variety of diagnostic purposes for a very long time. Similarly, relational databases have been employing commit logs to store information durably almost since their inception. However, developers' use of logs as a first-class persistence solution for structured information in mainstream systems remains extremely rare.

A log is an extremely simple, append-only sequence of immutable records ordered by time. The diagram here illustrates the structure of a log where records are written sequentially. In essence, a log is an append-only data structure, as depicted here:

Figure 5.9 – The log data structure

Writing to a log compared to a more complex data structure such as a table is a relatively simple and fast operation, and can handle extremely high volumes of data while providing predictable performance. Indeed, a modern event-streaming platform such as Apache Kafka makes use of this pattern to scale to support extremely high volumes. We do feel that this can be applied to act as a persistence store when processing commands in mainstream systems because this has benefits beyond the technical advantages listed previously. Consider the example of an online order flow shown here:

User action	Traditional store	Event store
Add milk to cart	Order 123: Milk in cart	E1: Cart#123 created
		E2: Milk added to cart
Add white bread to cart	Order 123: Milk and white bread in cart	E1: Cart#123 created
		E2: Milk added to cart
		E3: White bread added to cart
Remove white bread from cart	Order 123: Milk in cart	E1: Cart#123 created
		E2: Milk added to cart
		E3: White bread added to cart
		E4: White bread removed from cart
Add wheat bread to cart	Order 123: Milk and wheat bread in cart	E1: Cart#123 created
		E2: Milk added to cart
		E3: White bread added to cart
		E4: White bread removed from cart
		E5: Wheat bread added to cart
Confirm cart checkout	Order 123: Ordered milk and wheat bread	E1: Cart#123 created
		E2: Milk added to cart
		E3: White bread added to cart
		E4: White bread removed from cart
		E5: Wheat bread added to cart
		E6: Order 123 confirmed

As you can see, in the event store, we continue to have full visibility of all user actions performed. This allows us to reason about these behaviors more holistically. In the traditional store, we lost the information that the user replaced white with wheat bread. While this does not impact the order itself, we lose the opportunity to gather insights from this user behavior. We recognize that this information can be captured in other ways using specialized analytical solutions; however, the event log mechanism provides a natural way to do this without requiring any additional effort. It also acts as an audit log, providing a full history of all events that have occurred thus far. This fits well with the essence of domain-driven design, where we are constantly exploring ways in which to reduce complexity.

However, there are implications to persisting data in the form of a simple event log. Before processing any command, we will need to hydrate past events in the exact order of occurrence and reconstruct the aggregate state to allow us to perform validations. For example, when confirming checkout, just having the ordered set of elapsed events will not suffice. We still need to compute the exact items that are in the cart before allowing the order to be placed. This *event replay* to restore aggregate state (at least those attributes that are required to validate said command) is necessary before processing that command. For example, we need to know which items are in the cart currently before processing `RemoveItemFromCart Command`. This is illustrated in the following table:

Elapsed events	Aggregate state	Command	Event(s) emitted
—	—	Add item: milk	E1: Cart#123 created E2: Milk added
E1: Cart#123 created E2: Milk added	Cart items: Milk	Add item: white bread	E2: White bread added
E1: Cart#123 created E2: Milk added E3: White bread added	Cart items: Milk and white bread	Remove item: white bread	E3: White bread removed
E1: Cart#123 created E2: Milk added E3: White bread added E4: White bread removed	Cart items: Milk	Add item: wheat bread	E4: Wheat bread added
E1: Cart#123 created E2: Milk added E3: White bread added E4: White bread removed E5: Wheat bread added	Cart items: Milk and wheat bread	Confirm checkout for Cart#123	E5: Order created!

The corresponding source code for the whole scenario is illustrated in the following code snippet:

```
 1   public class Cart {
 2
 3       private boolean isNew;
 4       private CartItems items;
 5       //..
 6
 7       private Cart() {                                                      ❶
 8           // Required by the framework
 9       }
10
11       @CommandHandler
12       public void addItem(AddItemToCartCommand command) {
13           // Business validations here
14           if (this.isNew) {
15               apply(new CartCreatedEvent(command.getId()));             ❷
16           }
17           apply(new ItemAddedEvent(id, command.getItem()));            ❷
18       }
19
20       // Other command handlers omitted for brevity
21
22       @EventSourcingHandler                                                ❸
23       private void on(CartCreatedEvent event) {
24           this.id = event.getCartId();
25           this.items = new CartItems();
26           this.isNew = true;
27       }
28       // Other event sourcing handlers omitted for brevity
29   }
```

1. Before processing any command, the aggregate loading process commences by first invoking the no-args constructor. For this reason, we need the no-args constructor to be **empty**; that is, it should **not** have any code that restores state. Restoration of state *must* happen only in those methods that trigger an event replay. In the case of Axon Framework, this translates to methods embellished with the @EventSourcingHandler annotation.

2. It is important to note that it is possible (but not necessary) to emit **more than one event** after processing a command. This is illustrated in the command handler method for AddItemCommand in the previous code where we emit CartCreatedEvent and ItemAddedEvent. Command handlers do not mutate the state of the aggregate. They only make use of the existing aggregate state to enforce invariants (validations) and emit events if those invariants hold true.

3. The loading process continues through the invocation of event-sourcing handler methods in exactly the order of occurrence for that aggregate instance. Event-sourcing handlers are only needed to hydrate aggregate state on the basis of past events. This means that they usually are devoid of any business (conditional) logic. It goes without saying that these methods do not emit any events. Event emission is restricted to happen within command handlers when invariants are successfully enforced.

When working with event-sourced aggregates, it is very important to be disciplined about the kind of code that one can write:

Type of method	State restoration	Business logic	Event emission
@CommandHandler	No	Yes	Yes
@EventSourcingHandler	Yes	No	No

If there is a large number of historic events to restore state, the aggregate loading process can become a time-consuming operation – directly proportional to the number of elapsed events for that aggregate. There are techniques (such as event snapshotting) we can employ to overcome this.

Persistence technology choices

If you are using a state store to persist your aggregates, using your usual evaluation process for choosing your persistence technology should suffice. However, if you are looking at event-sourced aggregates, the decision can be a bit more nuanced. In our experience, even a simple relational database can do the trick. Indeed, we once made use of a relational database to act as an event store for a high-volume transactional application with billions of events. This setup worked just fine for us. It is worth noting that we were only using the event store to insert new events and loading events for a given aggregate in sequential order. However, there is a multitude of specialized technologies that have been purpose-built to act as an event store that supports several other value-added features, such as time travel, full event replay, event payload introspection, and so on. If you have such requirements, it might be worth considering other options, such as NoSQL databases (document stores such as MongoDB or column family stores such as Cassandra) or purpose-built commercial offerings, such as EventStoreDB and Axon Server, to evaluate feasibility in your context.

Which persistence mechanism should we choose?

Now that we have a reasonably good understanding of the two types of aggregate persistence mechanisms (state-stored and event-sourced), it begs the question of which one we should choose. We list a few benefits of using event sourcing here:

- We get to use the events as a **natural audit log** in high-compliance scenarios.
- It provides the ability to perform **more insightful analytics** on the basis of the fine-grained events data.
- It arguably produces more flexible designs when we work with a system based on **immutable events,** because the complexity of the persistence model is capped. Also, there is no need to deal with complex ORM impedance mismatch problems.
- The domain model is much more **loosely coupled** with the persistence model, enabling it to evolve mostly independently from the persistence model.

- It enables going back in time to be able to create **ad hoc views and reports** without having to deal with upfront complexity.

On the flip side, these are some challenges that you might have to consider when implementing an event-sourced solution:

- Event sourcing requires a **paradigm shift**, which means that development and business teams will have to spend time and effort understanding how it works.

- The persistence model does not store state directly. This means that **ad hoc querying** directly on the persistence model can be a lot more **challenging**. This can be alleviated by materializing new views; however, there is added complexity in doing that.

- Event sourcing usually tends to work very well when implemented in conjunction with CQRS, which arguably may add more complexity to the application. It also requires applications to pay closer attention to **strong versus eventual consistency** concerns.

Our experiences indicate that event-sourced systems bring a lot of benefits in modern event-driven systems. However, you will need to be cognizant of the considerations presented previously in the context of your own ecosystems when making persistence choices.

Enforcing policies

When processing commands, we need to enforce policies or rules. Policies come in two broad categories:

- Structural rules – those that enforce that the syntax of the dispatched command is valid

- Domain rules – those that enforce that business rules are adhered to

It may also be prudent to perform these validations in different layers of the system. And it is also common for some or all of these policy enforcements to be repeated in more than one layer of the system. However, the important thing to note is that before a command is successfully handled, all these policy enforcements are uniformly applied. Let's look at some examples of these in the upcoming section.

Structural validations

Currently, to create an LC application, you are required to dispatch `CreateLCApplicationCommand`. While the command dictates a structure, none of it is enforced at the moment. Let's correct that.

To be able to enable validations declaratively, we will make use of the JSR-303 bean validation libraries. We can add that easily, using the `spring-boot-starter-validation` dependency in our `pom.xml` file, as shown here:

```
1    <dependency>
2        <groupId>org.springframework.boot</groupId>
3        <artifactId>spring-boot-starter-validation</artifactId>
4    </dependency>
```

Now, we can add validations to the command object using the JSR-303 annotations, as depicted here:

```
1   import lombok.Data;
2   import javax.validation.*;
3   import javax.validation.constraints.*;
4
5   @Data
6   public class CreateLCApplicationCommand {
7
8       @NotNull
9       private LCApplicationId id;
10
11      @NotNull
12      private ClientId clientId;
13
14      @NotNull
15      @Valid
16      private Party applicant;
17
18      @NotNull
19      @Valid
20      private Party beneficiary;
21
22      @NotNull
23      @Valid
24      private AdvisingBank advisingBank;
25
26      @Future
27      private LocalDate issueDate;
28
29      @Positive
30      private MonetaryAmount amount;
31
32      @NotBlank
33      private String merchandiseDescription;
34  }
```

Most structural validations can be accomplished using built-in validator annotations. It is also possible to create custom validators for individual fields or to validate the entire object (for example, to validate interdependent attributes). For more details on how to do this, please refer to the bean validation specification at `https://beanvalidation.org/2.0/` and the reference implementation at `http://hibernate.org/validator/`.

Business rule enforcements

Structural validations can be accomplished using information that is already available in the command. However, there is another class of validations that requires information that is not present in the incoming command itself. This kind of information can be present in one of two places – within the aggregate that we are operating on or outside of the aggregate itself, but made available within the bounded context.

Let's look at an example of a validation that requires state to be present within the aggregate. Consider the example of submitting an LC. While we can make several edits to the LC when it is in a draft state, no changes can be made after it is submitted. This means that we can only submit an LC once. This act of submitting the LC is achieved by issuing `SubmitLCApplicationCommand`, as shown in the artifact from the eventstorming session:

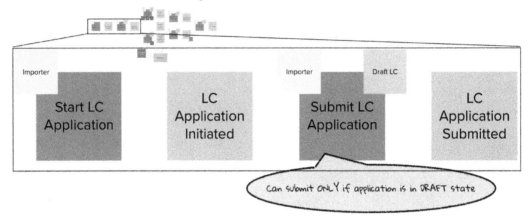

Figure 5.10 – Validations during the Submit LC Application process

Let's begin with a test to express our intent:

```
 1  class LCApplicationAggregateTests {
 2      //..
 3      @Test
 4      void shouldAllowSubmitOnlyInDraftState() {
 5          final LCApplicationId applicationId = LCApplicationId.randomId();
 6
 7          fixture.given(new LCApplicationCreatedEvent(applicationId))         ❶
 8                 .when(new SubmitLCApplicationCommand(applicationId))          ❷
 9                 .expectEvents(new LCApplicationSubmittedEvent(applicationId)); ❸
10      }
11  }
```

1. It is given that `LCApplicationCreatedEvent` has already occurred – in other words, the LC application is already created.

2. This is when we try to submit the application by issuing `SubmitLCApplicationCommand` for the same application.

3. We expect `LCApplicationSubmittedEvent` to be emitted.

The corresponding implementation will look something like the following:

```
1  class LCApplication {
2      // ..
3      @CommandHandler
4      public void submit(SubmitLCApplicationCommand command) {
5          apply(new LCApplicationSubmittedEvent(id));
6      }
7  }
```

The preceding implementation allows us to submit an LC application unconditionally – more than once. However, we want to restrict users to be able to submit only once. To be able to do that, we need to remember that the LC application has already been submitted. We can do that in the @ EventSourcingHandler handler of the corresponding events, as shown next:

```
1  class LCApplication {
2      // ..
3      @EventSourcingHandler
4      private void on(LCApplicationSubmittedEvent event) {
5          this.state = State.SUBMITTED;  ❶
6      }
7  }
```

1.   When LCApplicationSubmittedEvent is replayed, we set the state of the LC application to SUBMITTED.

While we have remembered that the application has changed to be in the SUBMITTED state, we are still not preventing more than one submit attempt. We can fix that by writing a test, as shown here:

```
1  class LCApplicationAggregateTests {
2      @Test
3      void shouldNotAllowSubmitOnAnAlreadySubmittedLC() {
4          final LCApplicationId applicationId = LCApplicationId.randomId();
5
6          fixture.given(
7                  new LCApplicationCreatedEvent(applicationId),            ❶
8                  new LCApplicationSubmittedEvent(applicationId))          ❶
9
10                 .when(new SubmitLCApplicationCommand(applicationId))     ❷
11
12                 .expectException(AlreadySubmittedException.class)        ❸
13                 .expectNoEvents();                                       ❹
14      }
15 }
```

1. `LCApplicationCreatedEvent` and `LCApplicationSubmittedEvent` have already happened, which means that `LCApplication` has been submitted once.

2. We now dispatch another `SubmitLCApplicationCommand` command to the system.

3. We expect `AlreadySubmittedException` to be thrown.

4. We also expect no events to be emitted.

The implementation of the command handler to make this work is shown here:

```
 1  class LCApplication {
 2      // ..
 3      @CommandHandler
 4      public void submit(SubmitLCApplicationCommand command) {
 5          if (this.state != State.DRAFT) {                                    ❶
 6              throw new AlreadySubmittedException("LC is already submitted!");
 7          }
 8          apply(new LCApplicationSubmittedEvent(id));
 9      }
10  }
```

1. Note how we are using the state attribute from the `LCApplication` aggregate to perform the validation. If the application is not in the DRAFT state, we fail with the `AlreadySubmittedException` domain exception.

Let's also look at an example where information needed to perform the validation is not part of either the command or the aggregate. Let's consider the scenario where country regulations prohibit transacting with a set of so-called *sanctioned* countries. Changes to this list of countries may be affected by external factors. Hence it does not make sense to pass this list of sanctioned countries as part of the command payload. Neither does it make sense to maintain it as part of every single aggregate's state – given that it can change (albeit very infrequently). In such a case, we may want to consider making use of a command handler that is outside the confines of the aggregate class. Thus far, we have only seen examples of `@CommandHandler` methods within the aggregate. But the `@CommandHandler` annotation can appear on any other class external to the aggregate. However, in such a case, we need to load the aggregate ourselves. Axon Framework provides an `org.axonframework.modelling.command.Repository` interface to allow us to do that. It is important to note that this repository is distinct from the Spring Framework interface that is part of the Spring data libraries. An example of how this works is shown here:

```
1   import org.axonframework.modelling.command.Repository;
2
3   class MyCustomCommandHandler {
4
5       private final Repository<LCApplication> repository;            ❶
6
7       MyCustomCommandHandler(Repository<LCApplication> repository) {
8           this.repository = repository;                             ❶
9       }
10
11      @CommandHandler
12      public void handle(SomeCommand command) {
13          Aggregate<LCApplication> application
14              = repository.load(command.getAggregateId());          ❷
15          // Command handling code
16      }
17
18      @CommandHandler
19      public void handle(AnotherCommand command) {
20          Aggregate<LCApplication> application
21              = repository.load(command.getAggregateId());
22          // Command handling code
23      }
24  }
```

1. We are injecting the Axon `Repository` interface to allow us to load aggregates. This was not required previously because the `@CommandHandler` annotation appeared on aggregate methods directly.

2. We are using the `Repository` interface to load aggregates and work with them. The `Repository` interface supports other convenient methods to work with aggregates. Please refer to the Axon Framework documentation for more usage examples.

Coming back to the sanctioned countries example, let's look at how we need to set up the test slightly differently:

```
1   public class CreateLCApplicationCommandHandlerTests {
2       private FixtureConfiguration<LCApplication> fixture;
3
4       @BeforeEach
5       void setUp() {
6           final Set<Country> sanctioned = Set.of(SOKOVIA);
7           fixture = new AggregateTestFixture<>(LCApplication.class);         ❶
8
9           final Repository<LCApplication> repository = fixture.getRepository();  ❷
10
11          CreateLCApplicationCommandHandler handler =
12              new CreateLCApplicationCommandHandler(repository, sanctioned);  ❸
13          fixture.registerAnnotatedCommandHandler(handler);                   ❹
14      }
15  }
```

1. We are creating a new aggregate fixture as usual.

2. We are using the fixture to obtain an instance of the Axon `Repository` interface.

3. We instantiate the custom command handler passing in the `Repository` instance. Also, note how we inject the collection of sanctioned countries into the handler using simple dependency injection. In real life, this set of sanctioned countries will likely be obtained from the external configuration.

4. We finally need to register the command handler with the fixture so that it can route commands to this handler as well.

The tests for this look fairly straightforward:

```
class CreateLCApplicationCommandHandlerTests {
    // ..

    @BeforeEach
    void setUp() {
    final Set<Country> sanctioned = Set.of(SOKOVIA);          ❶
        fixture = new AggregateTestFixture<>(LCApplication.class);

        final Repository<LCApplication> repository = fixture.getRepository();

        CreateLCApplicationCommandHandler handler =
                new CreateLCApplicationCommandHandler(repository, sanctioned);   ❷
        fixture.registerAnnotatedCommandHandler(handler);
    }

    @Test
    void shouldFailIfBeneficiaryCountryIsSanctioned() {
        fixture.given()
                .when(new CreateLCApplicationCommand(randomId(), SOKOVIA))        ❸
                .expectNoEvents()
                .expectException(CannotTradeWithSanctionedCountryException.class);
    }

    @Test
    void shouldCreateIfCountryIsNotSanctioned() {
        final LCApplicationId applicationId = randomId();
        fixture.given()
                .when(new CreateLCApplicationCommand(applicationId, WAKANDA))     ❹
                .expectEvents(new LCApplicationCreatedEvent(applicationId));
    }
}
```

1. For the purposes of the test, we mark the country SOKOVIA as a sanctioned country. In a more realistic scenario, this will likely come from some form of external configuration (for example, a lookup table or form of external configuration). However, this is appropriate for our unit test.

2. We then inject this set of sanctioned countries into the command handler.

3. When the LC application is created for the sanctioned country, we expect no events to be emitted and, furthermore, the `CannotTradeWithSanctionedCountryException` exception to be thrown.

4. Finally, when the beneficiary belongs to a non-sanctioned country, we emit `LCApplicationCreatedEvent` to be emitted.

The implementation of the command handler is shown here:

```
import org.springframework.stereotype.Service;

@Service                                                          ❶
public class CreateLCApplicationCommandHandler {
    private final Repository<LCApplication> repository;
    private final Set<Country> sanctionedCountries;

    public CreateLCApplicationCommandHandler(Repository<LCApplication> repository,
                                     Set<Country> sanctionedCountries) {
        this.repository = repository;
        this.sanctionedCountries = sanctionedCountries;
    }

    @CommandHandler
    public void handle(CreateLCApplicationCommand command) {
        // Validations can be performed here as well        ❷
        repository.newInstance(()
            -> new LCApplication(command, sanctionedCountries));  ❸
    }
}
```

1. We mark the class as `@Service` to mark it as a component devoid of encapsulated state and enable auto-discovery when using annotation-based configuration or classpath scanning. As such, it can be used to perform any "plumbing" activities.

2. Do note that the validation for the beneficiary's country being sanctioned could have been performed on line 18 as well. Some would argue that this would be ideal because we could avoid a potentially unnecessary invocation of the Axon `Repository` method if we did that. However, we prefer encapsulating business validations within the confines of the aggregate as much as possible, so that we don't suffer from the problem of creating an anemic domain model.

3. We use an aggregate repository to act as a factory to create a new instance of the `LCApplication` domain object.

Finally, the aggregate implementation along with the validation is shown here:

```
class LCApplication {
// ...
    public LCApplication(CreateLCApplicationCommand command, Set<Country> sanctioned) {
        if (sanctioned.contains(command.getBeneficiaryCountry())) { ❶
            throw new CannotTradeWithSanctionedCountryException();
        }
        apply(new LCApplicationCreatedEvent(command.getId()));
    }
}
```

1. The validation itself is fairly straightforward. We throw `CannotTradeWithSanctionedCountryException` when the validation fails.

With these examples, we looked at different ways to implement the policy enforcements encapsulated within the boundaries of the aggregate.

Summary

In this chapter, we used the outputs of the eventstorming session and used it as a primary aid to create a domain model for our bounded context. We looked at how to implement this using the CQRS architecture pattern. We looked at persistence options and the implications of using event-sourced versus state-stored aggregates. Finally, we rounded off by looking at a variety of ways in which to perform business validations. We looked at all this through a set of code examples, using Spring Boot and Axon Framework. With this knowledge, we should be able to implement robust, well-encapsulated, event-driven domain models.

In the next chapter, we will look at implementing a UI for these domain capabilities and examine a few options, such as CRUD-based versus task-based UIs.

Further reading

Title	Author	Location
CQRS	Martin Fowler	`https://martinfowler.com/bliki/CQRS.html`
Bootiful CQRS and Event Sourcing with Axon Framework	SpringDeveloper and Allard Buijze	`https://www.youtube.com/watch?v=7e5euKxHhTE`
The Log: What every software engineer should know about real-time data's unifying abstraction	Jay Kreps	`https://engineering.linkedin.com/distributed-systems/log-what-every-software-engineer-should-know-about-real-time-datas-unifying`
Event Sourcing	Martin Fowler	`https://martinfowler.com/eaaDev/EventSourcing.html`
Using a DDD Approach for Validating Business Rules	Fabian Lopez	`https://www.infoq.com/articles/ddd-business-rules/`
Anemic Domain Model	Martin Fowler	`https://www.martinfowler.com/bliki/AnemicDomainModel.html`

Implementing the User Interface – Task-Based

To accomplish a difficult task, one must first make it easy.

– Marty Rubin

The essence of **Domain-Driven Design (DDD)** is about capturing the business process and user intent. In the previous chapter, we designed a set of APIs without paying much attention to how those APIs would get consumed by their eventual users. In this chapter, we will design the GUI for the LC application using the JavaFX framework. As part of that, we will examine how this approach of designing APIs in isolation can cause an impedance mismatch between the producers and the consumers. We will examine the consequences of this impedance mismatch and how task-based UIs can help cope with this mismatch.

We will cover the following topics in this chapter:

- API styles
- Bootstrapping the UI
- Implementing the UI

By the end of the chapter, you will know how to employ DDD principles to help you build robust user experiences that are simple and intuitive. You will also learn why it may be prudent to design your backend interfaces (APIs) from the perspective of the consumer.

Technical requirements

You will need access to the following:

- JDK 1.8+ (we have used Java 16 to compile sample sources)
- JavaFX SDK 16 and Scene Builder

- Maven 3.x

- Spring Boot 2.4.x

- mvvmFX 1.8 (`https://sialcasa.github.io/mvvmFX/`)

- JUnit 5.7.x (included with Spring Boot)

- TestFX (for UI testing)

- OpenJFX Monocle (for headless UI testing)

- Project Lombok (to reduce verbosity)

Before we dive deep into building the GUI solution, let's do a quick recap of where we left the APIs.

API styles

If you recall from *Chapter 5, Implementing Domain Logic*, we created the following commands:

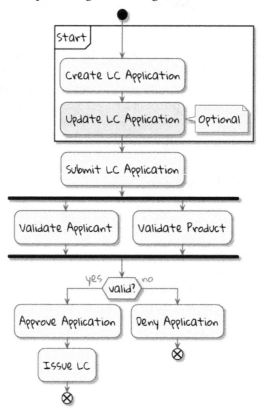

Figure 6.1 – Commands from the eventstorming session

If you observe carefully, there seem to be commands at two levels of granularity. **Create LC Application** and **Update LC Application** are coarse-grained, whereas the others are a lot more focused in terms of their intent. One possible decomposition of the coarse-grained commands can be as depicted here:

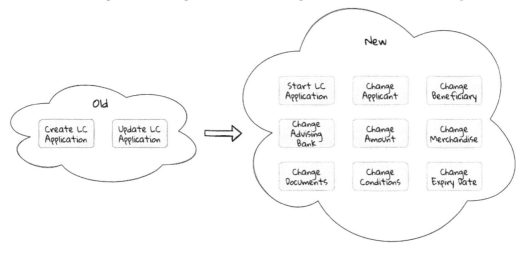

Figure 6.2 – Decomposed commands

In addition to just being more fine-grained than the commands in the previous iteration, the revised commands seem to better capture the user's intent. This may feel like a minor change in semantics, but can have a huge impact on the way our solution is used by its ultimate end users. The question then is whether we should *always* prefer fine-grained APIs over coarse-grained ones. The answer can be a lot more nuanced. When designing APIs and experiences, we see two main styles being employed:

- CRUD-based
- Task-based

Let's look at each of these in a bit more detail.

CRUD-based APIs

CRUD is an acronym used to refer to the four basic operations that can be performed on database applications: **create**, **read**, **update**, and **delete**. Many programming languages and protocols have their own equivalent of CRUD, often with slight variations in naming and intent. For example, SQL – a popular language for interacting with databases – calls the four functions `Insert`, `Select`, `Update`, and `Delete`. Similarly, the HTTP protocol has `POST`, `GET`, `PUT`, and `DELETE` as verbs to represent these CRUD operations. This approach has been extended to the design of our APIs as well. This has resulted in the proliferation of both CRUD-based APIs and user experiences. Take a look at `CreateLCApplicationCommand` from *Chapter 5, Implementing Domain Logic*:

```
 1   import lombok.Data;
 2
 3   @Data
 4   public class CreateLCApplicationCommand {
 5
 6       private LCApplicationId id;
 7
 8       private ClientId clientId;
 9       private Party applicant;
10       private Party beneficiary;
11       private AdvisingBank advisingBank;
12       private LocalDate issueDate;
13       private MonetaryAmount amount;
14       private String merchandiseDescription;
15   }
```

Along similar lines, it would not be uncommon to create a corresponding
UpdateLCApplicationCommand, as depicted here:

```
 1   import lombok.Data;
 2
 3   @Data
 4   public class UpdateLCApplicationCommand {
 5
 6       @TargetAggregateIdentifier
 7       private LCApplicationId id;
 8
 9       private ClientId clientId;
10       private Party applicant;
11       private Party beneficiary;
12       private AdvisingBank advisingBank;
13       private LocalDate issueDate;
14       private MonetaryAmount amount;
15       private String merchandiseDescription;
16   }
```

While this is very common and also very easy to grasp, it is not without problems. Here are some
questions that taking this approach raises:

- Are we allowed to change everything listed in the update command?

- Assuming that everything can change, do they all change at the same time?

- How do we know what exactly changed? Should we be doing a diff?

- What if all the attributes mentioned above are not included in the update command?

- What if we need to add attributes in the future?

- Is the business intent of what the user wanted to accomplish captured?

In a simple system, the answer to these questions may not matter that much. However, as system complexity increases, will this approach remain resilient to change? We feel that it merits taking a look at another approach called task-based APIs to be able to answer these questions.

Task-based APIs

In a typical organization, individuals perform tasks relevant to their specialization. The bigger the organization, the higher the degree of specialization. This approach of segregating tasks according to one's specialization makes sense because it mitigates the possibility of stepping on each other's shoes, especially when getting complex pieces of work done. For example, in the LC application process, there is a need to establish the value/legality of the product while also determining the creditworthiness of the applicant. It makes sense that each of these tasks is usually performed by individuals in unrelated departments. It also follows that these tasks can be performed independently of the other.

In terms of a business process, if we have a single `CreateLCApplicationCommand` command that precedes these operations, individuals in both departments firstly have to wait for the entire application to be filled out before either can commence their work. Secondly, if either piece of information is updated through a single `UpdateLCApplicationCommand` command, it is unclear what changed. This can result in a spurious notification being sent to at least one department because of this lack of clarity in the process.

Since most work happens in the form of specific tasks, it can work to our advantage if our processes and, by extension, our APIs mirror these behaviors.

Keeping this in mind, let's re-examine our revised APIs for the LC application process:

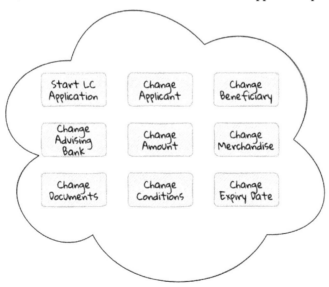

Figure 6.3 – Revised commands

While it may have appeared previously that we have simply converted our coarse-grained APIs to become more fine-grained, this, in reality, is a better representation of the tasks that the user intended to perform. So, in essence, task-based APIs are the decomposition of work in a manner that aligns more closely to the users' intents. With our new APIs, product validation can commence as soon as `ChangeMerchandise` happens. Also, it is unambiguously clear what the user did and what needs to happen in reaction to the user's action. It then begs the question of whether we should employ task-based APIs all the time. Let's look at the implications in more detail.

Task-based or CRUD-based?

CRUD-based APIs seem to operate at the level of the aggregate. In our example, we have the LC aggregate. In the simplest case, this essentially translates to four operations aligned with each of the CRUD verbs. However, as we are seeing, even in our simplified version, the LC is becoming a fairly complex concept. Having to work with just four operations at the level of the LC is cognitively complex. With more requirements, this complexity will only continue to increase. For example, consider a situation where the business expresses a need to capture a lot more information about `merchandise`, where today, this is simply captured in the form of free-form text. A more elaborate version of merchandise information is shown here:

```java
public class Merchandise {
    private MerchandiseId id;
    private Set<Item> items;
    private Packaging packaging;
    private boolean hazardous;
}

class Item {
    private ProductId productId;
    private int quantity;
    // ...
}

class Packaging {
    // ...
}
```

In our current design, the implications of this change are far-reaching for both the provider and the consumer(s). Let's look at some of the consequences in more detail:

Characteristic	CRUD-based	Task-based	Commentary
Usability	👎	👍	Task-based interfaces tend to provide more fine-grained controls that capture user intent a lot more explicitly, making them naturally more usable – especially in cases where the domain is complex.
Reusability	👎	👍	Task-based interfaces enable more complex features to be composed using simpler ones, providing more flexibility to the consumers.
Scalability	👎	👍	Task-based interfaces have an advantage because they can provide the ability to independently scale specific features. However, if the fine-grained task-based interfaces are used almost all the time in unison, it may be required to re-examine whether the users' intents are accurately captured.
Security	👎	👍	For task-based interfaces, security is enhanced from the producer's perspective by enabling the application of the *principle of least privilege*.
Latency	👍	👎	Arguably, coarse-grained CRUD interfaces can enable consumers to achieve a lot more in fewer interactions, thereby providing low latency.
Management Overhead	👍	👎	For the provider, fine-grained interfaces require a lot more work managing a larger number of interfaces.
Complexity	👎	👍	Complexity of the system as a whole is proportional to the number of features that need to be implemented. Assuming accidental complexity is avoided in both cases, task-based interfaces allow spreading complexity more or less uniformly across multiple simpler interfaces.

In our current design, the implications of this change are far-reaching for both the provider and the consumer(s). Let's look at some of the consequences in more detail.

As we can see, the decision between CRUD-based and task-based interfaces is nuanced. We are not suggesting that you should choose one over the other. Which style you use will depend on your specific requirements and context. In our experience, task-based interfaces treat user intents as first-class citizens and perpetuate the spirit of DDD's ubiquitous language very elegantly. Our preference is to design interfaces as task-based where possible because they result in more intuitive interfaces that better express the problem domain.

As systems evolve, and they begin to support richer user experiences and multiple channels, the CRUD-based interface seems to require additional translation layers to cater to user experience needs. The visual here depicts a typical layered architecture of a solution that supports multiple-user experience channels:

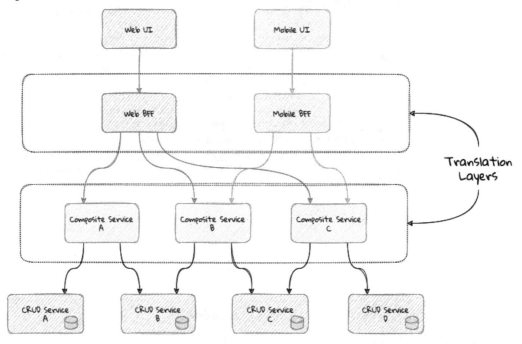

Figure 6.4 – Layered architecture supporting multiple-user experience channels

This setup is usually composed of the following:

- A domain tier comprised of CRUD-based services that simply map closely to database entities
- A composite tier comprised of business capabilities that span more than one core service
- A **Backend for Frontend** (**BFF**) tier, comprised of channel-specific APIs

Note that the composite and BFF tiers exist primarily as a means to map backend capabilities to user intent. In an ideal world, where backend APIs reflect user intent closely, the need for translations should be minimal (if at all). Our experience suggests that such a setup causes business logic to get pushed closer to the user channels as opposed to being encapsulated within the confines of well-factored business services. In addition, these tiers cause inconsistent experiences across channels for the same functionality, given that modern teams are structured along tier boundaries.

> **Important Note**
>
> We are not opposed to the use of layered architectures. We recognize that a layered architecture can bring modularity, separation of concerns, and other related benefits. However, we are opposed to creating additional tiers merely as a means to compensate for poorly factored core domain APIs.

A well-factored API tier can have a profound effect on how great user experiences are built. However, this is a chapter on implementing the user interface. Let's revert to creating the user interface for the LC application.

Bootstrapping the UI

We will simply be building the UI for the LC application we created in *Chapter 5, Implementing Domain Logic*. For detailed instructions, refer to the *Bootstrapping the application* section. In addition, we will need to add the following dependencies to the `dependencies` section of the Maven `pom.xml` file in the root directory of the project:

```
 1  <dependencies>
 2      <!--...-->
 3      <dependency>
 4          <groupId>org.openjfx</groupId>
 5          <artifactId>javafx-controls</artifactId>
 6          <version>${javafx.version}</version>
 7      </dependency>
 8      <dependency>
 9          <groupId>org.openjfx</groupId>
10          <artifactId>javafx-graphics</artifactId>
11          <version>${javafx.version}</version>
12      </dependency>
13      <dependency>
14          <groupId>org.openjfx</groupId>
15          <artifactId>javafx-fxml</artifactId>
16          <version>${javafx.version}</version>
17      </dependency>
18      <dependency>
19          <groupId>de.saxsys</groupId>
20          <artifactId>mvvmfx</artifactId>
21          <version>${mvvmfx.version}</version>
22      </dependency>
23      <dependency>
24          <groupId>de.saxsys</groupId>
25          <artifactId>mvvmfx-spring-boot</artifactId>
26          <version>${mvvmfx.version}</version>
27      </dependency>
28      <!--...-->
29  </dependencies>
```

To run UI tests, you will need to add the following dependencies:

```
1  <dependencies>
2      <!--...-->
3      <dependency>
4          <groupId>org.testfx</groupId>
5          <artifactId>testfx-junit5</artifactId>
6          <scope>test</scope>
7          <version>${testfx-junit5.version}</version>
8      </dependency>
9      <dependency>
10         <groupId>org.testfx</groupId>
11         <artifactId>openjfx-monocle</artifactId>
12         <version>${openjfx-monocle.version}</version>
13     </dependency>
14     <dependency>
15         <groupId>de.saxsys</groupId>
16         <artifactId>mvvmfx-testing-utils</artifactId>
17         <version>${mvvmfx.version}</version>
18         <scope>test</scope>
19     </dependency>
20     <!--...-->
21  </dependencies>
```

To be able to run the application from the command line, you will need to add `javafx-maven-plugin` to the `plugins` section of your `pom.xml` file, per the following:

```
1  <plugin>
2      <groupId>org.openjfx</groupId>
3      <artifactId>javafx-maven-plugin</artifactId>
4      <version>${javafx-maven-plugin.version}</version>
5      <configuration>
6          <mainClass>com.premonition.lc.ch06.App</mainClass>
7      </configuration>
8  </plugin>
```

To run the application from the command line, use the following:

```
mvn javafx:run
```

> **Important Note**
>
> If you are using a JDK greater than version 1.8, the JavaFX libraries may not be bundled with the JDK itself. When running the application from your IDE, you will likely need to add the following:
>
> ```
> --module-path=<path-to-javafx-sdk>/lib/ \
> --add-modules=javafx.controls,javafx.graphics,
> javafx.fxml,javafx.media
> ```

We are making use of the mvvmFX framework to assemble the UI. To make this work with Spring Boot, the application launcher looks as depicted here:

```
 1  @SpringBootApplication
 2  public class App extends MvvmfxSpringApplication {  ❶
 3
 4      public static void main(String[] args) {
 5          Application.launch(args);
 6      }
 7
 8      @Override
 9      public void startMvvmfx(Stage stage) {
10          stage.setTitle("LC Issuance");
11
12          final Parent parent = FluentViewLoader
13                  .fxmlView(MainView.class)
14                  .load().getView();
15
16          final Scene scene = new Scene(parent);
17          stage.setScene(scene);
18          stage.show();
19      }
20  }
```

> **Important Note**
>
> We are required to extend from the MvvmfxSpringApplication mvvmFX framework class.

Please refer to the ch06 directory of the accompanying source code repository for the complete example.

Implementing the UI

When working with user interfaces, it is fairly customary to use one of these presentation patterns:

- **Model View Controller (MVC)**
- **Model View Presenter (MVP)**
- **Model View View Model (MVVM)**

The MVC pattern has been around for the longest time. The idea of separating concerns among collaborating model, view, and controller objects is a sound one. However, beyond the definition of these objects, actual implementations seem to vary wildly – with the controller becoming overly complex in a lot of cases. In contrast, MVP and MVVM, while being derivatives of MVC, seem to bring about a better separation of concerns between the collaborating objects. MVVM, in particular when coupled with data-binding constructs, makes for code that is much more readable, maintainable, and testable. In this book, we make use of MVVM because it enables test-driven development, which is a strong personal preference for us. Let's look at a quick MVVM primer, as implemented in the mvvmFX framework.

MVVM primer

Modern UI frameworks started adopting a declarative style to express the view. MVVM was designed to remove all GUI code (code-behind) from the view by making use of binding expressions. This allowed for a cleaner separation of stylistic versus programming concerns. A high-level visual of how this pattern is implemented is shown here:

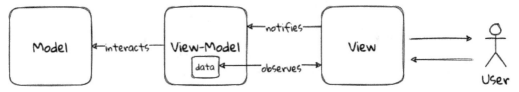

Figure 6.5 – MVVM design pattern

The pattern comprises the following components:

- **Model**: Responsible for housing the business logic and managing the state of the application.
- **View**: Responsible for presenting data to the user and notifying the view-model about user interactions through the view delegate.
- **View Delegate**: Responsible for keeping the view and the view-model in sync as changes are made by the user or on the view-model. It is also responsible for transmitting actions performed on the view to the view-model.
- **View-Model**: Responsible for handling user interactions on behalf of the view. The view-model interacts with the view using the observer pattern (typically, one-way or two-way data binding to make it more convenient). The view-model interacts with the model for updates and read operations.

Creating a new LC

Let's consider the example of creating a new LC. To start the creation of a new LC, all we need is for the applicant to provide a friendly client reference. This is an easy-to-remember string of free text. A simple rendition of this UI is shown here:

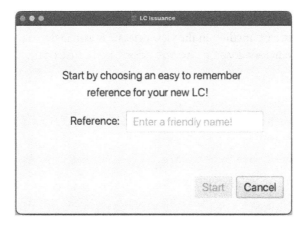

Figure 6.7 – Start LC creation screen

Let's examine the implementation and purpose of each component in more detail.

Declarative view

When working with JavaFX, the view can be rendered using a declarative style in FXML format. Important excerpts from the StartLCView.fxml file to start creating a new LC are shown here:

```
 1  <?import javafx.scene.layout.Pane?>
 2  <?import javafx.scene.control.Button?>
 3  <?import javafx.scene.control.TextField?>
 4
 5  <Pane id="start-lc"  xmlns="http://javafx.com/javafx/16"
 6                       xmlns:fx="http://javafx.com/fxml/1"
 7      fx:controller="com.premonition.lc.ch06.ui.views.StartLCView">   ❶
 8      ...
 9
10      <TextField id="client-reference"
11              fx:id="clientReference"/>                                ❷
12
13      <Button id="start-button"
14              fx:id="startButton"
15              text="Start"
16              onAction="#start"/>                                      ❸
17      ...
18  </Pane>
```

1. The StartLCView class acts as the view delegate for the FXML view and is assigned using the fx:controller attribute of the root element (javafx.scene.layout.Pane, in this case).

2. In order to reference the client-reference input field in the view delegate, we use the fx:id annotation (clientReference, in this case).

3. Similarly, the start button is referenced using `"fx:id=startButton"` in the view delegate. Furthermore, the `start` method in the view delegate is assigned to handle the default action (the button press event for `javafx.scene.control.Button`).

View delegate

Next, let's look at the structure of the `com.premonition.lc.issuance.ui.views.StartLCView` view delegate:

```
1   import javafx.fxml.FXML;
2   //...
3   public class StartLCView {                              ❶
4
5       @FXML
6       private TextField clientReference;                  ❷
7       @FXML
8       private Button startButton;                         ❸
9
10      public void start(ActionEvent event) {              ❹
11          // Handle button press logic here
12      }
13
14      // Other parts omitted for brevity...
15  }
```

1. This is the view delegate class for the `StartLCView.fxml` view.

2. This is the Java binding for the `clientReference` textbox in the view. The name of the member needs to match exactly with the value of the `fx:id` attribute in the view. Further, it needs to be annotated with the `@javafx.fxml.FXML` annotation. The use of the `@FXML` annotation is optional if the member in the view delegate is public and matches the name in the view.

3. Similarly, `startButton` is bound to the corresponding button widget in the view.

4. This is the method for the action handler when `startButton` is pressed.

View-model

The `StartLCViewModel` view-model class for `StartLCView` is shown here:

```
1   import javafx.beans.property.StringProperty;
2   import de.saxsys.mvvmfx.ViewModel;
3
4   public class StartLCViewModel implements ViewModel {        ❶
5
6       private final StringProperty clientReference;           ❷
7
8       public StartLCViewModel() {
9           this.clientReference = new SimpleStringProperty();  ❸
10      }
11
12      public StringProperty clientReferenceProperty() {       ❹
13          return clientReference;
14      }
15
16      public String getClientReference() {
17          return clientReference.get();
18      }
19
20      public void setClientReference(String clientReference) {
21          this.clientReference.set(clientReference);
22      }
23
24      // Other getters and setters omitted for brevity
25  }
```

1. This is the view-model class for `StartLCView`. Note that we are required to implement the `de.saxsys.mvvmfx.ViewModel` interface provided by the mvvmFX framework.

2. We are initializing the `clientReference` property using `SimpleStringProperty` provided by JavaFX. There are several other property classes to define more complex types. Please refer to the JavaFX documentation for more details.

3. This is the value of `clientReference` in the view-model. We will look at how to associate this with the value of the `clientReference` textbox in the view shortly. Note that we are using `StringProperty` provided by JavaFX, which provides access to the underlying string value of the client reference.

4. JavaFX beans are required to create a special accessor for the property itself in addition to the standard getter and setter for the underlying value.

Binding the view to the view-model

Next, let's look at how to associate the view to the view-model:

```
1   import de.saxsys.mvvmfx.Initialize;
2   import de.saxsys.mvvmfx.FxmlView;
3   import de.saxsys.mvvmfx.InjectViewModel;
4   //...
5   public class StartLCView implements FxmlView<StartLCViewModel> {      ❶
6
7       @FXML
8       private TextField clientReference;
9       @FXML
10      private Button startButton;
11
12      @InjectViewModel
13      private StartLCViewModel viewModel;                                ❷
14
15      @Initialize
16      private void initialize() {                                        ❸
17          clientReference.textProperty()
18              .bindBidirectional(viewModel.clientReferenceProperty());   ❹
19          startButton.disableProperty()
20              .bind(viewModel.startDisabledProperty());                  ❺
21      }
22
23      // Other parts omitted for brevity...
24  }
```

1. The mvvmFX framework requires that the `view delegate` implement the FXMLView<? extends ViewModelType>. In this case, the view-model type is StartLCViewModel. The mvvmFX framework supports other view types as well. Please refer to the framework documentation for more details.

2. The framework provides a @de.saxsys.mvvmfx.InjectViewModel annotation to allow dependency, injecting the view-model into the view delegate.

3. The framework will invoke all methods annotated with the @de.saxsys.mvvmfx.Initialize annotation during the initialization process. The annotation can be omitted if the method is named initialize and is declared public. Please refer to the framework documentation for more details.

4. We have now bound the text property of the clientReference textbox in the view delegate to the corresponding property in the view-model. Note that this is a bidirectional binding, which means that the value in the view and the view-model are kept in sync if it changes on either side.

5. This is another variation of binding in action, where we are making use of a unidirectional binding. Here, we are binding the disabled property of the start button to the corresponding property on the view-model. We will look at why we need to do this shortly.

Enforcing business validations in the UI

We have a business validation that the client reference for an LC needs to be at least four characters in length. This will be enforced on the backend. However, to provide a richer user experience, we will also enforce this validation on the UI.

> **Important Note**
>
> This may feel contrary to the notion of centralizing business validations on the backend. While this may be a noble attempt at implementing the **Don't Repeat Yourself** (**DRY**) principle, in reality, it poses a lot of practical problems. Distributed systems expert Udi Dahan has a very interesting take on why this may not be such a virtuous thing to pursue. Ted Neward also talks about this in his blog, entitled *The Fallacies of Enterprise Computing*.

The advantage of using MVVM is that this logic is easily testable in a simple unit test of the view-model. Let's see this in action now:

```
class StartLCViewModelTests {

    private StartLCViewModel viewModel;

    @BeforeEach
    void before() {
        int clientReferenceMinLength = 4;
        viewModel = new StartLCViewModel(clientReferenceMinLength);
    }

    @Test
    void shouldNotEnableStartByDefault() {
        assertThat(viewModel.getStartDisabled()).isTrue();
    }

    @Test
    void shouldNotEnableStartIfClientReferenceLesserThanMinimumLength() {
        viewModel.setClientReference("123");
        assertThat(viewModel.getStartDisabled()).isTrue();
    }

    @Test
    void shouldEnableStartIfClientReferenceEqualToMinimumLength() {
        viewModel.setClientReference("1234");
        assertThat(viewModel.getStartDisabled()).isFalse();
    }

    @Test
    void shouldEnableStartIfClientReferenceGreaterThanMinimumLength() {
        viewModel.setClientReference("12345");
        assertThat(viewModel.getStartDisabled()).isFalse();
    }
}
```

Now, let's look at the implementation for this functionality in the view-model:

```
 1  public class StartLCViewModel implements ViewModel {
 2
 3      //...
 4      private final StringProperty clientReference;
 5      private final BooleanProperty startDisabled;            ❶
 6
 7      public StartLCViewModel(int clientReferenceMinLength) {  ❷
 8          this.clientReference = new SimpleStringProperty();
 9          this.startDisabled = new SimpleBooleanProperty();
10          this.startDisabled
11              .bind(this.clientReference.length()
12                      .lessThan(clientReferenceMinLength));   ❸
13      }
14
15      //...
16  }
17
18  public class StartLCView implements FxmlView<StartLCViewModel> {
19
20      //...
21      @Initialize
22      public void initialize() {
23          startButton.disableProperty()
24              .bind(viewModel.startDisabledProperty());       ❹
25          clientReference.textProperty()
26              .bindBidirectional(viewModel.clientReferenceProperty());
27      }
28      //...
29  }
```

1. We declare a `startDisabled` property in the view-model to manage when the start button should be disabled.

2. The minimum length for a valid client reference is injected into the view-model. It is conceivable that this value will be provided as part of the external configuration, or possibly from the backend.

3. We create a binding expression to match the business requirement.

4. We bind the view-model property to the disabled property of the start button in the view delegate.

Let's also look at how to write an end-to-end, headless UI test, as shown here:

```
1   @UITest
2   public class StartLCViewTests {                                              ①
3
4       @Autowired
5       private ApplicationContext context;
6
7       @Init
8       public void init() {
9           MvvmFX.setCustomDependencyInjector(context::getBean);              ②
10      }
11
12      @Start
13      public void start(Stage stage) {                                        ③
14          final Parent parent = FluentViewLoader
15                  .fxmlView(StartLCView.class)
16                  .load().getView();
17          stage.setScene(new Scene(parent));
18          stage.show();
19      }
20
21      @Test
22      void blankClientReference(FxRobot robot) {
23          robot.lookup("#client-reference")                                   ④
24              .queryAs(TextField.class)
25              .setText("");
26
27          verifyThat("#start-button", NodeMatchers.isDisabled());             ⑤
28      }
29
30      @Test
31      void validClientReference(FxRobot robot) {
32          robot.lookup("#client-reference")
33              .queryAs(TextField.class)
34              .setText("Test");
35
36          verifyThat("#start-button", NodeMatchers.isEnabled());              ⑤
37      }
38  }
39
```

1. We have written a convenience @UITest extension to combine the Spring Framework and TestFX testing. Please refer to the accompanying source code with the book for more details.

2. We set up the Spring context to act as the dependency injection provider for the mvvmFX framework and its injection annotations (for example, @InjectViewModel) to work.

3. We are using the @Start annotation provided by the TestFX framework to launch the UI.

4. The TestFX framework injects an instance of the FxRobot UI helper, which we can use to access UI elements.

5. We are using the TestFX framework-provided convenience-matchers for test assertions.

Now, when we run the application, we can see that the start button is enabled when a valid client reference is entered:

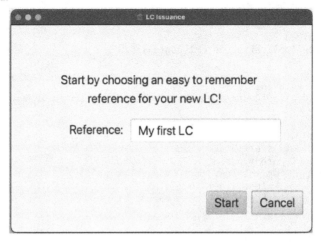

Figure 6.8 – The start button is enabled with a valid client reference

Now that we have the start button enabling correctly, let's implement the actual creation of the LC itself by invoking the backend API.

Integrating with the backend

LC creation is a complex process, requiring information about a variety of items as evidenced when we decomposed the LC creation process. In this section, we will integrate the UI with the command to start the creation of a new LC. This happens when we press the **Start** button on the **Start LC** creation screen. The revised `StartNewLCApplicationCommand` looks as shown here:

```
1   @Data
2   public class StartNewLCApplicationCommand {
3       private final String applicantId;
4       private final LCApplicationId id;
5       private final String clientReference;
6
7       private StartNewLCApplicationCommand(String applicantId, String clientReference) {
8           this.id = LCApplicationId.randomId();
9           this.applicantId = applicantId;
10          this.clientReference = clientReference;
11      }
12
13      public static StartNewLCApplicationCommand startApplication( ❶
14                      String applicantId,
15                      String clientReference) {
16          return new StartNewLCApplicationCommand(applicantId, clientReference);
17      }
18  }
```

1. To start a new LC application, we need `applicantId` and `clientReference`.

Given that we are using the MVVM pattern, the code to invoke the backend service is part of the view-model. Let's test-drive this functionality:

```
1  @ExtendWith(MockitoExtension.class)
2  class StartLCViewModelTests {
3
4      @Mock
5      private BackendService service;
6
7      @BeforeEach
8      void before() {
9          int clientReferenceMinLength = 4;
10         viewModel = new StartLCViewModel(clientReferenceMinLength, service);
11     }
12
13     @Test
14     void shouldNotInvokeBackendIfStartButtonIsDisabled() {
15         viewModel.setClientReference("");
16         viewModel.startNewLC();
17
18         Mockito.verifyNoInteractions(service);
19     }
20 }
```

The view-model is enhanced accordingly to inject an instance of `BackendService` and looks as shown here:

```
1  public class StartLCViewModel implements ViewModel {
2
3      private final BackendService service;
4      // Other members omitted for brevity
5
6      public StartLCViewModel(int clientReferenceMinLength,
7                              BackendService service) {
8          this.service = service;
9          // Other code omitted for brevity
10     }
11
12     public void startNewLC() {
13         // TODO: invoke backend!
14     }
15 }
```

Now, let's test to actually make sure that the backend gets invoked only when a valid client reference is input:

```
1   class StartLCViewModelTests {
2       // ...
3
4       @BeforeEach
5       void before() {
6           viewModel = new StartLCViewModel(4, service);
7           viewModel.setLoggedInUser(new LoggedInUserScope("test-applicant"));    ❶
8       }
9
10      @Test
11      void shouldNotInvokeBackendIfStartButtonIsDisabled() {
12          viewModel.setClientReference("");
13          viewModel.startNewLC();
14
15          Mockito.verifyNoInteractions(service);                                 ❷
16      }
17
18      @Test
19      void shouldInvokeBackendWhenStartingCreationOfNewLC() {
20          viewModel.setClientReference("My first LC");
21          viewModel.startNewLC();
22
23          Mockito.verify(service).startNewLC("test-applicant", "My first LC");   ❸
24      }
25  }
```

1. We set the logged-in user.

2. When the client reference is blank, there should be no interactions with the backend service.

3. When a valid value for the client reference is entered, the backend should be invoked with the entered value.

The implementation to make this test pass looks like this:

```
1   public class StartLCViewModel {
2       //...
3       public void startNewLC() {
4           if (!getStartDisabled()) {                          ❶
5               service.startNewLC(
6                   userScope.getLoggedInUserId(),
7                   getClientReference());                      ❷
8           }
9       }
10      //...
11  }
```

1. We check that the start button is enabled before invoking the backend.

2. These are the actual backend calls with the appropriate values.

Now, let's look at how to integrate the backend call from the view. We test this in a UI test, as shown here:

```
1   @UITest
2   public class StartLCViewTests {
3
4       @MockBean
5       private BackendService service;                                    ❶
6
7       //...
8
9       @Test
10      void shouldLaunchLCDetailsWhenCreationIsSuccessful(FxRobot robot) {
11          final String clientReference = "My first LC";
12          LCApplicationId lcApplicationId = LCApplicationId.randomId();
13
14          when(service.startNewLC("test-applicant", clientReference))
15              .thenReturn(lcApplicationId);                              ❷
16
17          robot.lookup("#client-reference")
18              .queryAs(TextField.class)
19              .setText(clientReference);                                 ❸
20          robot.clickOn("#start-button");                                ❹
21
22          Mockito.verify(service).startNewLC(
23              "test-applicant", clientReference);                        ❺
24
25          verifyThat("#lc-details-screen", isVisible());                 ❻
26      }
27  }
```

1. We inject a mock instance of the backend service.

2. We stub the call to the backend to return successfully.

3. We type in a valid value for the client reference.

4. We click on the **Start** button.

5. We verify that the service was indeed invoked with the correct arguments.

6. We verify that we have moved to the next screen in the UI (the LC details screen).

Let's also look at what happens when the service invocation fails in another test:

```
1   public class StartLCViewTests {
2       //...
3       @Test
4       void shouldStayOnCreateLCScreenOnCreationFailure(FxRobot robot) {
5           final String clientReference = "My first LC";
6           when(service.startNewLC("test-applicant", clientReference))
7               .thenThrow(new RuntimeException("Failed!!"));             ❶
8
9           robot.lookup("#client-reference")
10              .queryAs(TextField.class)
11              .setText(clientReference);
12          robot.clickOn("#start-button");
13
14          verifyThat("#start-lc-screen", isVisible());                 ❷
15      }
16  }
```

1. We stub the backend service call to fail with an exception.

2. We verify that we continue to remain on `start-lc-screen`.

The view implementation for this functionality is shown here:

```
 1  import javafx.concurrent.Service;
 2
 3  public class StartLCView {
 4      //...
 5      public void start(ActionEvent event) {
 6          new Service<Void>() {                          1
 7              @Override
 8              private Task<Void> createTask() {
 9                  return new Task<>() {
10                      @Override
11                      private Void call() {
12                          viewModel.startNewLC();        2
13                          return null;
14                      }
15                  };
16              }
17
18              @Override
19              private void succeeded() {
20                  Stage stage = UIUtils.getStage(event);
21                  showLCDetailsView(stage);              3
22              }
23
24              @Override
25              private void failed() {
26                  // Nothing for now. Remain on the same screen.
27              }
28          }.start();
29      }
30  }
```

1. JavaFX, like most frontend frameworks, is single-threaded and requires that long-running tasks not be invoked on the UI thread. For this purpose, it provides the `javafx.concurrent.Service` abstraction to handle such interactions elegantly in a background thread.

2. The actual invocation of the backend through the view-model happens here.

3. We show the next screen to enter more LC details here.

Finally, the service implementation itself is shown here:

```
 1  import org.axonframework.commandhandling.gateway.CommandGateway;
 2
 3  @Service
 4  public class BackendService {
 5
 6      private final CommandGateway gateway;                          ❶
 7
 8      public BackendService(CommandGateway gateway) {
 9          this.gateway = gateway;
10      }
11
12      public LCApplicationId startNewLC(String applicantId, String clientReference) {
13          return gateway.sendAndWait(                                 ❷
14                  startApplication(applicantId, clientReference)
15              );
16      }
17  }
```

1. We inject `org.axonframework.commandhandling.gateway.CommandGateway` provided by Axon Framework to invoke the command.

2. The actual invocation of the backend using the `sendAndWait` method happens here. In this case, we are blocking until the backend call completes. There are other variations that do not require this kind of blocking. Please refer to the Axon Framework documentation for more details.

We have now seen a complete example of how to implement the UI and invoke the backend API.

Summary

In this chapter, we looked at the nuances of API styles and clarified why it is very important to design APIs that capture the users' intent closely. We looked at the differences between CRUD-based and task-based APIs. Finally, we implemented the UI making use of the MVVM design pattern and demonstrated how it aids in test-driving frontend functionality.

Now that we have implemented the creation of a new LC, for implementing the subsequent commands we will require access to an existing LC. In the next chapter, we will look at how to implement the query side and how to keep it in sync with the command side.

Further reading

Title	Author	Location
Task-driven user interfaces	Oleksandr Sukholeyster	https://www.uxmatters.com/mt/archives/2014/12/task-driven-user-interfaces.php
Business logic, a different perspective	Udi Dahan	https://vimeo.com/131757759
The Fallacies of Enterprise Computing	Ted Neward	http://blogs.tedneward.com/post/enterprise-computing-fallacies/
GUI architectures	Martin Fowler	https://martinfowler.com/eaaDev/uiArchs.html

7

Implementing Queries

The best view comes after the hardest climb.

– Anonymous

In the section *Command Query Responsibility Segregation (CQRS)* from *Chapter 3, Understanding the Domain*, we described how DDD and CQRS complement each other and how the query side (read models) can be used to create one or more representations of the underlying data. In this chapter, we will dive deeper into how we can construct read-optimized representations of the data by listening to domain events. We will also look at persistence options for these read models.

When working with query models, we construct models by listening to events as they happen. We will examine how to deal with the following situations:

- New requirements evolving over a period of time, requiring us to build new query models.

- We discover a bug in our query model that requires us to recreate the model from scratch.

To do that, the agenda of the chapter includes the following topics:

- Continuing our design journey

- Implementing the query side

- Historic event replays

By the end of this chapter, you will learn to appreciate how to build query models by listening to domain events. You will also learn how to purpose-build new query models to suit specific read requirements as opposed to being restricted by the data model that was chosen to service commands. You will finally look at how historic event replays work and how you can use them to create new query models to service new requirements.

Technical requirements

To follow the examples in this chapter, you will need access to the following:

- JDK 1.8+ (we have used Java 17 to compile sample sources)

- Spring Boot 2.4.x

- Axon Framework 4.5.3

- JUnit 5.7.x (included with Spring Boot)

- OpenJFX Monocle (for headless UI testing)

- Project Lombok (to reduce verbosity)

- Maven 3.x

Please refer to the `Chapter07` directory of the book's accompanying source code repository at `https://github.com/PacktPublishing/Domain-Driven-Design-with-Java-A-Practitioner-s-Guide/tree/master/Chapter07` for complete working examples.

Continuing our design journey

In *Chapter 4, Domain Analysis and Modeling*, we discussed eventstorming as a lightweight method to clarify business flows. As a reminder, this is the output produced from our eventstorming session:

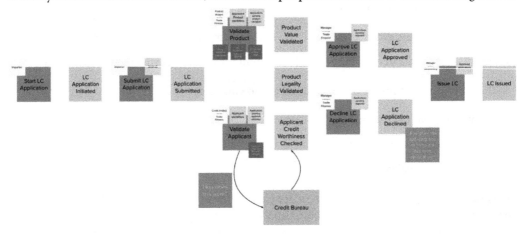

Figure 7.1 – Recap of the eventstorming session

As mentioned previously, we are making use of the CQRS architecture pattern to create the solution. For a detailed explanation on why this is a sound method to employ, you can revisit the *When to use CQRS* section in *Chapter 3, Understanding the Domain,* where we've already covered this. In the preceding diagram, the green stickies represent read/query models. These query models are required when validating a command (for example, a list of valid product identifiers when processing the `ValidateProduct` command) or if the information is simply required to be presented to the user (for example, a list of LCs created by an applicant). Let's look at what it means to apply CQRS in practical terms for the query side.

Implementing the query side

In *Chapter 5, Implementing Domain Logic*, we examined how to publish events when a command is successfully processed. Now, let's look at how we can construct a query model by listening to these domain events. Logically, this will look something like the following diagram:

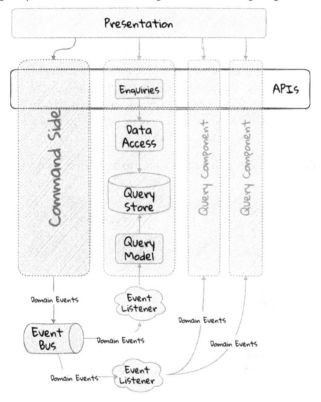

Figure 7.2 – The CQRS application – the query side

For a detailed explanation of how the command side is implemented, please refer to the *Implementing the command side* section in *Chapter 5, Implementing Domain Logic*.

The high-level sequence on the query side is described here:

1. An event listening component listens to these domain events published on the event bus.
2. It constructs a purpose-built query model to satisfy a specific query use case.
3. This query model is persisted in a datastore optimized for read operations.
4. This query model is then exposed in the form of an API.

Note how there can exist more than one query side component for handling respective scenarios.

Let's implement each of these steps to see how this works for our LC issuance application.

Tooling choices

In a CQRS application, there is a separation between the command and query sides. At this time, this separation is logical in our application because both the command and query sides are running as components within the same application process. To illustrate the concepts, we will use conveniences provided by Axon Framework to implement the query side in this chapter. In *Chapter 10, Beginning the Decomposition Journey*, we will look at how it may not be necessary to use a specialized framework (such as Axon) to implement the query side.

When implementing the query side, we have two concerns to address, as depicted in the following diagram:

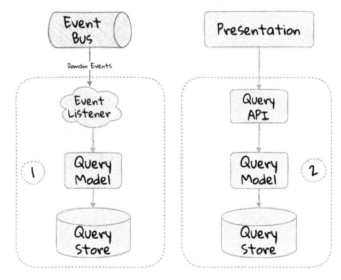

Figure 7.3 – The query side dissected

These concerns are as follows:

- Consuming domain events and persisting one or more query models
- Exposing the query model as an API

Before we start implementing these concerns, let's identify the queries we need to implement for our LC issuance application.

Identifying queries

From the eventstorming session, we have the following queries to start with:

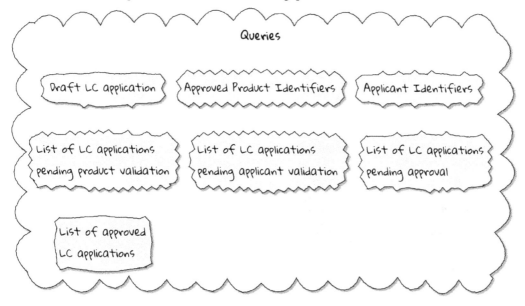

Figure 7.4 – Identified queries

The queries marked in green (in the output from the eventstorming session from *Figure 7.1*) all require us to expose a collection of LCs in various states. To represent this, we can create an LCView class, an extremely simple object devoid of any logic, as shown here:

```
public class LCView {

    private LCApplicationId id;
    private String applicantId;
    private String clientReference;
    private LCState state;

    // Getters and setters omitted for brevity
}
```

These query models are an absolute necessity to implement basic functionality dictated by business requirements. But it is possible and very likely that we will need additional query models as the system requirements evolve. We will enhance our application to support these queries as and when the need arises.

Creating the query model

As seen in *Chapter 5, Implementing Domain Logic*, when starting a new LC application, the importer sends `StartNewLCApplicationCommand`, which results in `LCApplicationStartedEvent` being emitted, as shown here:

```
 1 | class LCApplication {
 2 |     //..
 3 |     @CommandHandler
 4 |     public LCApplication(StartNewLCApplicationCommand command) {
 5 |         // Validation code omitted for brevity
 6 |         // Refer to chapter 5 for details.
 7 |         AggregateLifecycle.apply(new LCApplicationStartedEvent(command.getId(),
 8 |                 command.getApplicantId(), command.getClientReference()));
 9 |     }
10 |     //..
11 | }
```

Let's write an event-processing component that will listen to this event and construct a query model. When working with Axon Framework, we have a convenient way to do this by annotating the event-listening method with the `@EventHandler` annotation.

To make any method an event listener, we annotate it with the `@EventHandler` annotation:

```
 1 | import org.axonframework.eventhandling.EventHandler;
 2 | import org.springframework.stereotype.Component;
 3 |
 4 | @Component
 5 | class LCApplicationStartedEventHandler {
 6 |
 7 |     @EventHandler                                              ❶
 8 |     public void on(LCApplicationStartedEvent event) {
 9 |         LCView view = new LCView(event.getId(),
10 |                       event.getApplicantId(),
11 |                       event.getClientReference(),
12 |                       event.getState());                       ❷
13 |         // Perform any transformations to optimize access
14 |         repository.save(view);                                 ❸
15 |     }
16 | }
```

1. To make any method an event listener, we annotate it with the `@EventHandler` annotation.

2. The handler method needs to specify the event that we intend to listen to. There are other arguments that are supported by event handlers. Please refer to the Axon Framework documentation for more information.

3. We finally save the query model in an appropriate query store. When persisting this data, we should consider storing it in a form that is optimized for data access. In other words, we want to reduce as much complexity and cognitive load when querying this data.

The `@EventHandler` annotation should not be confused with the `@EventSourcingHandler` annotation that we looked at in *Chapter 5, Implementing Domain Logic*. The `@EventSourcingHandler` annotation is used to replay events and restore aggregate state when loading event-sourced aggregates on the command side, whereas the `@EventHandler` annotation is used to listen to events outside the context of the aggregate. In other words, the `@EventSourcingHandler` annotation is used exclusively within aggregates, whereas the `@EventHandler` annotation can be used anywhere there is a need to consume domain events. In this case, we are using it to construct a query model.

Query-side persistence choices

Segregating the query side this way enables us to choose the persistence technology most appropriate for the problem being solved on the query side. For example, if extreme performance and simple filtering criteria are important, it may be prudent to choose an in-memory store such as Redis or Memcached. If complex search/analytics requirements and large datasets are to be supported, then we may want to consider something like Elasticsearch. Or we may even simply choose to stick with just a relational database. The point we would like to emphasize is that employing CQRS affords a level of flexibility that was previously not available to us.

Exposing a query API

Applicants like to view the LCs they created, specifically those in the draft state. Let's look at how we can implement this functionality. Let's start by defining a simple object to capture the query criteria:

```
import org.springframework.data.domain.Pageable;

public class MyDraftLCsQuery {

    private ApplicantId applicantId;
    private Pageable page;

    // Getters and setters omitted for brevity
}
```

Let's implement the query using Spring's repository pattern to retrieve the results for these criteria:

```
1   import org.axonframework.queryhandling.QueryHandler;
2
3   public interface LCViewRepository extends JpaRepository<LCView,
4   LCApplicationId> {
5
6       Page<LCView> findByApplicantIdAndState(              ❶
7               String applicantId,
8               LCState state,
9               Pageable page);
10
11      @QueryHandler                                         ❷
12      default Page<LCView> on(MyDraftLCsQuery query) {
13          return findByApplicantIdAndState(                 ❸
14                  query.getApplicantId(),
15                  LCState.DRAFT,
16                  query.getPage());
17      }
    }
```

1. This is the dynamic Spring data finder method we will use to query the database.

2. The `@QueryHandler` annotation provided by Axon Framework routes query requests to the respective handler.

3. Finally, we invoke the finder method to return results.

In the preceding example, we have implemented the `QueryHandler` method within the repository itself for brevity. `QueryHandler` can be placed elsewhere as well.

To connect this to the UI, we add a new method in `BackendService` (originally introduced in *Chapter 6, Implementing the User Interface - Task-Based*) to invoke the query, as shown here:

```
1   import org.axonframework.queryhandling.QueryGateway;
2
3   public class BackendService {
4
5       private final QueryGateway queryGateway;              ❶
6
7       public List<LCView> findMyDraftLCs(String applicantId) {
8           return queryGateway.query(                        ❷
9                   new MyDraftLCsQuery(applicantId),
10                      ResponseTypes.multipleInstancesOf(LCView.class))
11                  .join();
12
13      }
14  }
```

1. Axon Framework provides the `QueryGateway` convenience that allows us to invoke the query. For more details on how to use `QueryGateway`, please refer to the Axon Framework documentation.

2. We execute the query using the `MyDraftLCsQuery` object to return results.

What we looked at previously is an example of a very simple query implementation where we have a single `@QueryHandler` annotation to service the query results. This implementation returns results as a one-time fetch. Let's look at more complex query scenarios.

Advanced query scenarios

Our focus currently is on active LC applications. Maintaining issued LCs happens in a different bounded context of the system. Consider a scenario where we need to provide a consolidated view of currently active LC applications and issued LCs. In such a scenario, it is necessary to obtain this information by querying two distinct sources (ideally in parallel) – commonly referred to as the *scatter-gather* pattern. Please refer to the section on scatter-gather queries in the Axon Framework documentation for more details.

In other cases, we may want to remain up to date on dynamically changing data. For example, consider a real-time stock-ticker application tracking price changes. One way to implement this is by polling for price changes. A more efficient way to do this is to push price changes as and when they occur – commonly referred to as the *publish-subscribe* pattern. Please refer to the section on subscription queries in the Axon Framework documentation for more details.

Historic event replays

The example we have looked at thus far allows us to listen to events as they occur. Consider a scenario where we need to build a new query from historic events to satisfy an unanticipated new requirement. This new requirement may necessitate the need to create a new query model or, in a more extreme case, a completely new bounded context. Another scenario might be when we may need to correct a bug in the way we had built an existing query model and now need to recreate it from scratch. Given that we have a record of all events that have transpired in the event store, we can use replay events to enable us to construct both new and/or correct existing query models with relative ease.

> **Important Note**
>
> We have used the term *event replay* in the context of reconstituting the state of event-sourced aggregate instances (discussed in the *Event-sourced aggregates* section in *Chapter 5, Implementing Domain Logic*). The event replay mentioned here, although similar in concept, is still very different. In the case of a domain object event replay, we work with a single aggregate root instance and only load events for that one instance. In this case, though, we will likely work with events that span more than one aggregate.

Let's look at the different types of replays and how we can use each of them.

Types of replays

When replaying events, there are at least two types of replays, depending on the requirements we need to meet. Let's look at each type in turn:

- **Full event replay**: This is where we replay all the events in the event store. This can be used in a scenario where we need to support a completely new bounded context that is dependent on this subdomain. This can also be used in cases where we need to support a completely new query model or reconstruct an existing, erroneously built query model. Depending on the number of events in the event store, this can be a fairly long and complex process.

- **Partial/ad hoc event replay**: This is where we need to replay all the events on a subset of aggregate instances or a subset of events on all aggregate instances, or a combination of both. When working with partial event replays, we will need to specify filtering criteria to select subsets of aggregate instances and events. This means that the event store needs to have the flexibility to support these use cases. Using specialized event store solutions (such as Axon Server and EventStoreDB, to name a couple) can be extremely beneficial.

Event replay considerations

The ability to replay events and create new query models can be invaluable. However, like everything else, there are considerations that we need to keep in mind when working with replays. Let's examine some of these in more detail.

Event store design

As mentioned in *Chapter 5*, *Implementing Domain Logic*, when working with event-sourced aggregates, we persist immutable events in the persistence store. The primary use cases that we need to support are as follows:

- Provide consistent and predictable **write** performance when acting as an append-only store.
- Provide consistent and predictable **read** performance when querying for events using the aggregate identifier.

However, replays (especially partial/ad hoc) require the event store to support much richer querying capabilities. Consider a scenario where we found an issue where the amount is incorrectly reported for LCs that were approved during a certain time period and only for a certain currency. To fix this issue, we need to do the following:

1. Identify affected LCs from the event store.
2. Fix the issue in the application.
3. Reset the query store for these affected aggregates.
4. Do a replay of a subset of events for the affected aggregates and reconstruct the query model.

Identifying affected aggregates from the event store can be tricky if we don't support querying capabilities that allow us to introspect the event payload. Even if this kind of ad hoc querying were to be supported, these queries can adversely impact the command-handling performance of the event store. One of the primary reasons to employ CQRS was to make use of query-side stores for such complex read scenarios.

Event replays seem to introduce a chicken and egg problem where the query store has an issue that can only be corrected by querying the event store. A few options to mitigate this issue are discussed here:

- **General purpose store**: Choose an event store that offers predictable performance for both scenarios (command handling and replay querying).

- **Built-in datastore replication**: Make use of read replicas for event replay querying.

- **Distinct datastores**: Make use of two distinct datastores to solve each problem on its own (for example, use a relational database/key-value store for command handling and a search-optimized document store for event replay querying).

> **Important Note**
> Do note that the distinct datastores approach for replays is used to satisfy an operational problem as opposed to query-side business use cases discussed earlier in this chapter. Arguably, it is more complex because the technology team on the command side has to be equipped to maintain more than one database technology.

Event design

Event replays are required to reconstitute the state from an event stream. In this article on what it means to be event-driven (https://martinfowler.com/articles/201701-event-driven.html), Martin Fowler talks about three different styles of events. If we employ the event-carried state-transfer approach (in Martin's article) to reconstitute state, it might require us to only replay the latest event for a given aggregate, as opposed to replaying all the events for that aggregate in order of occurrence. While this may seem convenient, it also has its downsides:

- All events may now require carrying a lot of additional information that may not be relevant to that event. Assembling all this information when publishing the event can add to the cognitive complexity on the command side.

- The amount of data that needs to be stored and flow through the wire can increase drastically.

- On the query side, it can increase cognitive complexity when understanding the structure of the event and processing it.

In a lot of ways, this leads back to the CRUD-based versus task-based approach for APIs discussed in *Chapter 5*, *Implementing Domain Logic*. Our general preference is to design events with as lean a payload as possible. However, your experiences may be different, depending on your specific problem or situation.

Application availability

In an event-driven system, it is common to accumulate an extremely large number of events over a period of time, even in a relatively simple application. Replaying a large number of events can be time-consuming. Let's look at the mechanics of how replays typically work:

1. We suspend listening to new events in preparation for a replay.
2. Clear the query store for impacted aggregates.
3. Start an event replay for impacted aggregates.
4. Resume listening to new events after the replay is complete.

Based on the preceding list, while the replay is running (*step 3*), we may not be able to provide reliable answers to queries that are impacted by the replay. This obviously has an impact on application availability. When using event replays, care needs to be taken to ensure that **Service Level Objectives (SLOs)** continue to be met.

Event handlers with side effects

When replaying events, we re-trigger event handlers either to fix logic that was previously erroneous or to support new functionality. Invoking most (if not all) event handlers usually results in some sort of side effect (for example, updating a query store). This means that some event handlers may not be running for the first time. To prevent unwanted side effects, it is important to undo the effects of having invoked these event handlers previously or code event handlers in an idempotent manner (for example, by using an `upsert` command instead of a simple `insert` command or an `update` command). The effect of some event handlers can be hard (if not impossible) to undo (for example, invoking a command, sending an email, or SMS). In such cases, it might be required to mark such event handlers as being ineligible to run during the replay. When using Axon Framework, this is fairly simple to do:

```
import org.axonframework.eventhandling.DisallowReplay;

class LCApplicationEventHandlers {
    @EventHandler
    @DisallowReplay    ❶
    public void on(CardIssuedEvent event) {
        // Behavior that we don't want replayed
    }
}
```

`@DisallowReplay` (or its counterpart `@AllowReplay`) can be used to explicitly mark event handlers ineligible to run during the replay.

Events as an API

In an event-sourced system where events are persisted instead of domain state, it is natural for the structure of events to evolve over a period of time. Consider an example of `BeneficiaryInformationChangedEvent`, which has evolved over a period of time, as shown here:

Figure 7.5 – Event evolution

Given that the event store is immutable, it is conceivable that we may have one or more combinations of these event versions for a given LC. This can present a number of decisions we will need to make when performing an event replay:

- The producer can simply provide the historic event as it exists in the event store and allow consumers to deal with older versions of the event.

- The producer can upgrade older versions of events to the latest version before exposing them to the consumer.

- Allow the consumer to specify an explicit version of the event that they are able to work with and upgrade it to that version before exposing it to the consumer.

- Migrate the events in the event store to the latest version as evolutions occur. This may not be feasible, given the immutable promise of events in the event store.

Which approach you choose really depends on your specific context and the maturity of the producer/consumer ecosystem. Axon Framework makes provisions for a process they call event upcasting that allows events to be upgraded just in time before they are consumed. Please refer to the Axon Framework documentation for more details.

In an event-driven system, events are your API. This means that you will need to apply the same rigor that you apply to APIs when making life cycle management decisions (for example, versioning, deprecation, and backward compatibility).

Summary

In this chapter, we examined how to implement the query side of a CQRS-based system. We looked at how domain events can be consumed in real time to construct materialized views that can be used to service query APIs. We looked at the different query types that can be used to efficiently access the underlying query models. We rounded off by looking at persistence options for the query side.

Finally, we looked at historic event replays and how they can be used to correct errors or introduce new functionality in an event-driven system.

This chapter should give you a good idea of how to build and evolve the query side of a CQRS-based system to meet changing business requirements while retaining all the business logic on the command side.

In this chapter, we looked at how to consume events in a stateless manner (where no two event handlers have knowledge of each other's existence). In the next chapter, we will continue to look at how to consume events, but this time in a stateful manner, in the form of long-running user transactions (also known as sagas).

Further reading

Title	Location
Scatter gather pattern	`https://www.enterpriseintegrationpatterns.com/BroadcastAggregate.html`
Publish-subscribe pattern	`https://www.enterpriseintegrationpatterns.com/PublishSubscribeChannel.html`
Axon Server	`https://axoniq.io/product-overview/axon`
EventStoreDB	`https://www.eventstore.com/eventstoredb`
Styles of events	`https://martinfowler.com/articles/201701-event-driven.html`
Event upcasting	`https://docs.axoniq.io/reference-guide/axon-framework/events/event-versioning#event-upcasting`
Service level objectives	`https://sre.google/sre-book/service-level-objectives`

8

Implementing
Long-Running Workflows

In the long run, the pessimist may be proven right, but the optimist has a better time on the trip.

— *Daniel Reardon*

In the previous chapters, we looked at handling commands and queries within the context of a single aggregate. All the scenarios we have looked at thus far have been limited to a single interaction. However, not all capabilities can be implemented in the form of a simple request-response interaction, requiring coordination across multiple external systems or human-centric operations, or both. In other cases, there may be a need to react to triggers that are non-deterministic (occur conditionally or not at all) and/or are time-bound (based on a deadline). This may require managing business transactions across multiple bounded contexts that may run over a long duration of time while continuing to maintain consistency (**saga**).

There are at least two common patterns to implement the saga pattern:

- **Explicit orchestration**: A designated component acts as a centralized coordinator—where the system relies on the coordinator to react to domain events to manage the flow.

- **Implicit choreography**: No single component is required to act as a centralized coordinator— where the components simply react to domain events in other components to manage the flow.

We'll cover the following main topics in this chapter:

- Implementing sagas

- Deciding between orchestration and choreography

- Handling deadlines

By the end of this chapter, you will have learned how to implement sagas using both techniques. You will also have learned how to work with deadlines when no explicit events occur within the system. You will finally be able to appreciate when/whether to choose an explicit orchestrator or simply stick to implicit choreography without resorting to the use of potentially expensive distributed transactions.

Technical requirements

To follow the examples in this chapter, you will need access to the following:

- JDK 1.8+ (we have used Java 17 to compile sample sources)

- Spring Boot 2.4.x

- Axon Framework 4.5.3

- JUnit 5.7.x (included with Spring Boot)

- Project Lombok (to reduce verbosity)

- Maven 3.x

Please refer to the `Chapter08` directory of the book's accompanying source code repository for complete working examples on GitHub at `https://github.com/PacktPublishing/Domain-Driven-Design-with-Java-A-Practitioner-s-Guide/tree/master/Chapter08`.

Continuing our design journey

In *Chapter 4, Domain Analysis and Modeling*, we discussed eventstorming as a lightweight method to clarify business flows. As a reminder, this is the output produced from our eventstorming session:

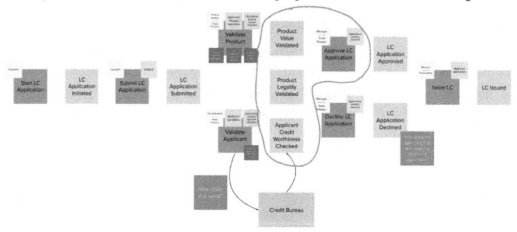

Figure 8.1 – Recap of the eventstorming session

As depicted in the preceding diagram, some aspects of **Letter of Credit** (**LC**) application processing happen outside our current bounded context, before the trade finance manager makes a decision to either approve or decline the application, as listed here:

- Product value is validated.

- Product legality is validated.

- Applicant's creditworthiness is validated.

Currently, the final approval is a manual process. It is pertinent to note that the product value and legality checks happen as part of the work done by the product analysis department, whereas applicant creditworthiness checks happen in the credit analysis department. Both departments make use of their own systems to perform these functions and notify us through the respective events. An LC application is **not ready** to be either approved or declined until **each** of these checks is completed. Each of these processes happens mostly independently of the others and may take a non-deterministic amount of time (typically in the order of a few days). After these checks have happened, the trade finance manager manually reviews the application and makes the final decision.

Given the growing volumes of LC applications received, the bank is looking to introduce a process optimization to automatically approve applications with an amount below a certain threshold (*USD 10,000* at this time). The business has deemed that the three preceding checks are sufficient and that no further human intervention is required when approving such applications.

From an overall system perspective, it is pertinent to note that the product analyst system notifies us through `ProductValueValidatedEvent` and `ProductLegalityValidatedEvent`, whereas the credit analyst system does the same through the `ApplicantCreditValidatedEvent` event. Each of these events can, and indeed do, happen independently of the others. For us to be able to auto-approve applications, our solution needs to wait for all of these events to occur. Once these events have occurred, we need to examine the outcome of each of these events to finally make a decision.

> **Note**
>
> In this context, we are using the term long-running to denote a complex business process that takes several steps to complete. As these steps occur, the process transitions from one state to another. In other words, we are referring to a state machine. This is not to be confused with a long-running software process (for example, a complex SQL query or an image-processing routine) that is computationally intensive.

As is evident from the preceding diagram, the LC auto-approval functionality is an example of a long-running business process where *something* in our system needs to keep track of the fact that these independent events have occurred before proceeding further. Such functionality can be implemented using the saga pattern. Let's look at how we can do this.

Implementing sagas

Before we delve into how we can implement this auto-approval functionality, let's take a look at how this works from a logical perspective, as shown here:

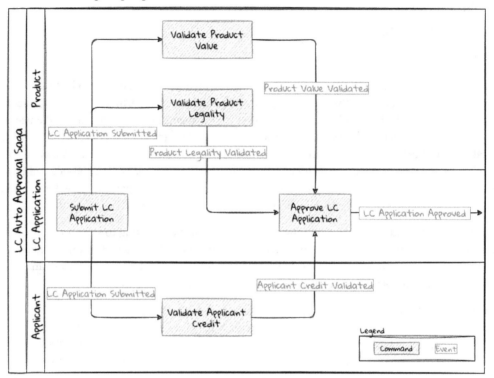

Figure 8.2 – Auto-approval process—logical view

As depicted in the preceding diagram, there are three bounded contexts in play:

- **LC Application** (the bounded context we have been implementing thus far)
- The **Applicant** bounded context
- The **Product** bounded context

The flow gets triggered when the LC application is submitted. This in turn sets in motion three independent functions that establish the following:

- Value of the product being transacted
- Legality of the product being transacted
- Creditworthiness of the applicant

LC approval can proceed only after **all** of these functions have been completed. Furthermore, to **auto-approve**, all of these checks have to complete **favorably** and, as mentioned earlier, the LC amount has to be less than the *USD 10,000* threshold.

As shown in the eventstorming artifact, the `LC Application` aggregate is able to handle `ApproveLCApplicationCommand`, which results in `LCApplicationApprovedEvent`. To auto-approve, this command needs to be invoked automatically when all the conditions mentioned earlier are satisfied. We are building an event-driven system, and we can see that each of these validations produces events when their respective actions are complete. There are at least two ways to implement this functionality:

- **Orchestration**: Where a single component in the system coordinates the state of the flow and triggers subsequent actions as necessary

- **Choreography**: Where actions in the flow are triggered without requiring an explicit coordinating component

Let's examine these methods in more detail.

Orchestration

When implementing sagas using an orchestrating component, the system looks similar to the one depicted here:

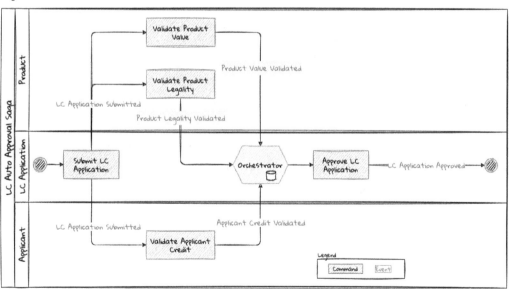

Figure 8.3 – Saga implementation using an orchestrator

The orchestrator starts tracking the flow when the LC application is submitted. It will then need to wait for each of the `ProductValueValidatedEvent`, `ProductLegalityValidatedEvent`, and `ApplicantCreditValidatedEvent` events to occur and decide whether it is appropriate to trigger `ApproveLCApplicationCommand`. Finally, the saga life cycle ends unconditionally when the LC application is approved. There are other conditions that may cause the saga to end abruptly. We will examine those scenarios in detail later. It is pertinent to note that there will be a **distinct** auto-approval saga instance for each LC application that gets submitted. Let's look at how to implement this functionality using the Axon Framework. As usual, let's test-drive this functionality that a new auto-approval saga instance is created when an LC application is submitted:

```java
import org.axonframework.test.saga.FixtureConfiguration;
import org.axonframework.test.saga.SagaTestFixture;

class AutoApprovalSagaTests {

    private FixtureConfiguration fixture;                              ❶

    @BeforeEach
    void setUp() {
        fixture = new SagaTestFixture<>(AutoApprovalSaga.class);       ❶
    }

    @Test
    void shouldStartSagaOnSubmit() {
        final LCApplicationId lcApplicationId = LCApplicationId.randomId();
        fixture.givenNoPriorActivity()                                 ❷
                .whenPublishingA(                                      ❸
                        new LCApplicationSubmittedEvent(lcApplicationId,
                                AUTO_APPROVAL_THRESHOLD_AMOUNT
                                    .subtract(ONE_DOLLAR)))
                .expectActiveSagas(1);                                 ❹
    }
}
```

1. We make use of the Axon-provided `FixtureConfiguration` and `SagaTestFixture`, which allow us to test saga functionality.

2. Given no prior activity has occurred (from the perspective of the saga).

3. When `LCApplicationSubmittedEvent` is published.

4. We expect one active saga to exist.

The implementation to make this test pass looks as follows:

```
1  import org.axonframework.modelling.saga.SagaEventHandler;
2  import org.axonframework.modelling.saga.StartSaga;
3  import org.axonframework.spring.stereotype.Saga;
4
5  @Saga                                                          ❶
6  public class AutoApprovalSaga {
7
8      @SagaEventHandler(associationProperty = "lcApplicationId") ❷
9      @StartSaga                                                 ❸
10      public void on(LCApplicationSubmittedEvent event) {
11          //
12      }
13 }
```

1. When working with Axon and Spring, the orchestrator is annotated with the @Saga annotation to mark it as a Spring bean. In order to track each submitted LC application, the @Saga annotation is prototype-scoped (as opposed to singleton-scoped), to allow the creation of multiple saga instances. Please refer to the Axon and Spring documentation for more information.

2. The saga listens to LCApplicationSubmittedEvent to keep track of the flow (as denoted by the @SagaEventHandler annotation). Conceptually, the @SagaEventHandler annotation is very similar to the @EventHandler annotation that we discussed in the previous chapter. However, the @SagaEventHandler annotation is used specifically for event listeners within a saga. The associationProperty attribute on the @SagaEventHandler annotation causes this event handler method to get invoked only for the saga with a matching value of the lcApplicationId attribute in the event payload. Also, @SagaEventHandler is a transaction boundary. Every time such a method completes successfully, the Axon Framework commits a transaction, thereby allowing it to keep track of the state stored in the saga. We will look at this in more detail shortly.

3. Every saga needs to have at least one @SagaEventHandler method that is also annotated with the @StartSaga annotation to denote the beginning of the saga.

We have a requirement that an LC cannot be auto-approved if its amount exceeds the threshold (*USD 10,000* in our case). The test for this scenario looks like this:

```
1  class AutoApprovalSagaTests {
2      //...
3
4      @Test
5      void shouldEndSagaImmediatelyIfAmountGreaterThanAutoApprovalThreshold() {
6          final LCApplicationId lcApplicationId = LCApplicationId.randomId();
7          fixture.givenAggregate(lcApplicationId.toString()).published()
8                  .whenPublishingA(
9                          new LCApplicationSubmittedEvent(lcApplicationId,
10                             AUTO_APPROVAL_THRESHOLD_AMOUNT.add(ONE_DOLLAR)))  ❶
11                  .expectActiveSagas(0);                                       ❷
12      }
13 }
```

1. When the LC amount exceeds the auto-approval threshold amount.

2. We expect no active sagas to exist for that LC.

The implementation to satisfy this condition looks like this:

```
import org.axonframework.modelling.saga.SagaLifecycle;

@Saga
public class AutoApprovalSaga {

    @SagaEventHandler(associationProperty = "lcApplicationId")
    @StartSaga
    public void on(LCApplicationSubmittedEvent event) {
        if (AUTO_APPROVAL_THRESHOLD_AMOUNT.isLessThan(event.getAmount())) { ❶
            SagaLifecycle.end();                                            ❷
        }
    }
}
```

1. We check for the condition of the LC amount being greater than the threshold amount.

2. If so, we end the saga using the framework-provided SagaLifecycle.end() method. Here, we end the saga programmatically. It is also possible to declaratively end the saga as well using the @EndSaga annotation when LCApplicationApprovedEvent occurs. Please refer to the full code examples included in this chapter's repository for more information.

We need to auto-approve the saga if ApplicantCreditValidatedEvent, ProductLegalityValidatedEvent, and ProductValueValidatedEvent have all occurred successfully. The test to verify this functionality is shown here:

```
class AutoApprovalSagaTests {

    @Test
    void shouldAutoApprove() {
        // Initialization code removed for brevity

        fixture.givenAggregate(lcApplicationId.toString())
            .published(submitted, legalityValidated, valueValidated)    ❶
                .whenPublishingA(applicantValidated)                    ❷
                .expectActiveSagas(1)                                   ❸
                .expectDispatchedCommands(
                        new ApproveLCApplicationCommand(lcApplicationId)); ❹
    }
}
```

1. Given that the LC application has been submitted and ProductValueValidatedEvent and ProductLegalityValidatedEvent have occurred successfully.

2. When ApplicantCreditValidatedEvent is published.

3. We expect one active saga instance and the following.

4. We expect ApproveLCApplicationCommand to be dispatched for that LC.

The implementation for this looks as follows:

```
1  class AutoApprovalSaga {
2
3      private boolean productValueValidated;                          ❶
4      private boolean productLegalityValidated;                       ❶
5      private boolean applicantValidated;                             ❶
6
7      @Autowired
8      private transient CommandGateway gateway;                       ❷
9
10     // Other event handlers omitted for brevity
11
12     @SagaEventHandler(associationProperty = "lcApplicationId")
13     public void on(ApplicantCreditValidatedEvent event) {           ❸
14         if (event.getDecision().isRejected()) {                     ❹
15             SagaLifecycle.end();
16         } else {
17             this.applicantValidated = true;                         ❺
18             if (productValueValidated && productLegalityValidated) { ❻
19                 LCApplicationId id = event.getLcApplicationId();
20                 gateway.send(ApproveLCApplicationCommand.with(id));  ❼
21             }
22         }
23     }
24
25     // Other event handlers omitted for brevity
26 }
```

1. As mentioned previously, sagas can maintain state. In this case, we are maintaining three Boolean variables, each to denote the occurrence of the respective event.

2. We have declared the Axon CommandGateway as a transient member because we need it to dispatch commands, but not be persisted along with other saga states.

3. This event handler intercepts ApplicantCreditValidatedEvent for the specific LC application (as denoted by associationProperty in the @SagaEventHandler annotation).

4. If the decision from ApplicantCreditValidatedEvent is rejected, we end the saga immediately.

5. Otherwise, we remember the fact that the applicant's credit has been validated.

6. We then check to see whether the product's value and legality have already been validated.

7. If so, we issue the command to auto-approve the LC.

> **Note**
>
> The logic in ProductValueValidatedEvent and ProductLegalityValidatedEvent is very similar to that in the saga event handler for ApplicantCreditValidatedEvent. We have omitted it here for brevity. Please refer to the source code for this chapter for the full example along with the tests.

Finally, we can end the saga when we receive `LCApplicationApprovedEvent` for this application:

```
class AutoApprovalSagaTests {
    @Test
    @DisplayName("should end saga after auto approval")
    void shouldEndSagaAfterAutoApproval() {
        // Initialization code omitted for brevity

        fixture.givenAggregate(lcApplicationId.toString())
                .published(
                    submitted, applicantValidated,
                    legalityValidated, valueValidated)           ❶
                .whenPublishingA(new LCApplicationApprovedEvent(lcApplicationId))  ❷
                .expectActiveSagas(0)                            ❸
                .expectNoDispatchedCommands();                   ❹
    }
}
```

1. Given that the LC has been submitted and all the validations have been completed successfully.

2. When `LCApplicationApprovedEvent` is published.

3. We expect zero active sagas to be running.

4. We also expect to not dispatch any commands.

Now that we have looked at how to implement sagas using an orchestrator, let's examine some design decisions that we may need to consider when working with them.

Here are the pros of orchestration:

- **Complex workflows**: Having an explicit orchestrator can be very helpful when dealing with flows that involve multiple participants and have a lot of conditionals because the orchestrator can keep track of the overall progress in a fine-grained manner.

- **Testing**: As we have seen in the preceding implementation, testing flow logic in isolation is relatively straightforward.

- **Debugging**: Given that we have a single coordinator, debugging the current state of the flow can be relatively easier.

- **Handling exceptions**: Given that the orchestrator has fine-grained control of the flow, recovering gracefully from exceptions can be easier.

- **System knowledge**: Components in different bounded contexts do not need to have knowledge of each other's internals (for example, commands and events) to progress the flow.

- **Cyclic dependencies**: Having a central coordinator allows avoiding accidental cyclic dependencies between components.

Here are the cons of orchestration:

- **Single point of failure**: From an operational perspective, orchestrators can become single points of failure because they are the only ones that have knowledge of the flow. This means that these components need to exhibit higher-resilience characteristics as compared to other components.

- **Leaking of domain logic**: In an ideal world, the aggregate will remain the custodian of all domain logic. Given that the orchestrator is also stateful, business logic may inadvertently shift to the orchestrator. Care should be taken to ensure that the orchestrator only has flow control logic while business invariants remain within the confines of the aggregate.

The preceding implementation should give you a good idea of how to implement a saga orchestrator. Now let's look at how we can do this without the use of an explicit orchestrator.

Choreography

Saga orchestrators keep track of the current state of the flow, usually making use of some kind of data store. Another way to implement this functionality is without using any stateful component. Logically, this looks like the setup shown in the diagram here:

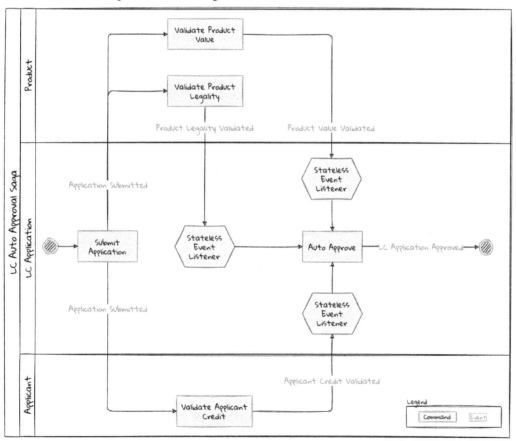

Figure 8.4 – Saga implementation using choreography

As you can see, there is no single component that tracks the saga life cycle. However, to make the auto-approval decision, each of these stateless event handlers needs to have knowledge of the same three events occurring:

- Product value is validated.

- Product legality is validated.

- Applicant's creditworthiness is validated.

Given that the event listeners themselves are stateless, there are at least three ways to provide this information to them:

- Each of the events carries this information in their respective payloads.

- The event listeners query the source systems (in this case, the product- and applicant-bounded contexts, respectively).

- The LC application-bounded context maintains a query model to keep track of these events occurring.

Just like in the orchestrator example, when all events have occurred and the LC amount is below the specified threshold, these event listeners can issue `ApproveLCApplicationCommand`.

> **Note**
> We will skip covering code examples for the choreography implementation because this is no different from the material we have covered previously in this and prior chapters.

Now that we have looked at how to implement both styles of sagas, let's examine some design decisions that we may need to consider when working with them.

The pros of the choreography implementation are listed here:

- **Simple workflows**: For simple flows, the choreography approach can be relatively straightforward because it does not require the overhead of an additional coordinating component.

- **No single points of failure**: From an operational perspective, there is one less high-resilience component to worry about.

These are the cons of the choreography implementation:

- **Workflow tracking**: Especially with complex workflows that involve numerous steps and conditionals, tracking and debugging the current state of the flow may become challenging.

- **Cyclic dependencies**: It is possible to inadvertently introduce cyclic dependencies among components when workflows become gnarly.

Sagas enable applications to maintain data and transactional consistency when more than one bounded context is required to complete the business functionality without having to resort to using *distributed transactions*. However, it does introduce a level of complexity to the programming model, especially when it comes to handling failures. We will look at exception handling in a lot more detail when we discuss working with distributed systems in upcoming chapters. Let's look at how to progress flows when there are no explicit stimuli by looking at how deadlines work.

Handling deadlines

Thus far, we have looked at events that are caused by human (for example, the applicant submitting an LC application) or system (for example, the auto-approval of an LC application) action. However, in an event-driven system, not all events occur due to an explicit human or system stimulus. Events may need to be emitted either due to inactivity over a period of time or on a recurring schedule based on prevailing conditions.

For example, let's examine the case where the bank needs *submitted LC applications* to be decided on as quickly as possible. When applications are not acted upon by the trade finance managers within 10 calendar days, the system should send them reminders.

To deal with such inactivity, we need a means by which to trigger system actions (read: emit events) based on the passage of time—in other words, perform actions when a *deadline* expires. In a happy path scenario, we expect either the user or the system to take a certain action. In such cases, we will also need to account for cases where we will need to cancel the trigger scheduled to occur on deadline expiry. Let's look at how to test-drive this functionality:

```
class LCApplicationAggregateTests {
    //...
    @Test
    void shouldCreateSubmissionReminderDeadlineWhenApplicationIsSubmitted() {
        final LCApplicationId id = LCApplicationId.randomId();
        fixture.given(new LCApplicationStartedEvent(id, ApplicantId.randomId(),
                            "My LC", LCState.DRAFT),
                    new LCAmountChangedEvent(id, THOUSAND_DOLLARS),
                    new MerchandiseChangedEvent(id, merchandise()))

                .when(new SubmitLCApplicationCommand(id)) ❶
                .expectEvents(new LCApplicationSubmittedEvent(id,
                            THOUSAND_DOLLARS))

                .expectScheduledDeadlineWithName(
                        Duration.ofDays(10),
                        LC_APPROVAL_PENDING_REMINDER); ❷
    }
}
```

1. When the LC application is submitted.

2. We expect a deadline for the reminder to be scheduled.

The implementation for this is fairly straightforward:

```
 1  import org.axonframework.deadline.DeadlineManager;
 2
 3  class LCApplication {
 4      //...
 5      @CommandHandler
 6      public void on(SubmitLCApplicationCommand command,
 7                      DeadlineManager deadlineManager) {  ❶
 8          assertPositive(amount);
 9          assertMerchandise(merchandise);
10          assertInDraft(state);
11          apply(new LCApplicationSubmittedEvent(id, amount));
12
13          deadlineManager.schedule(Duration.ofDays(10),  ❷
14              "LC_APPROVAL_REMINDER",
15              LCApprovalPendingNotification.first(id));  ❸
16      }
17      //...
18  }
```

1. To allow working with deadlines, the Axon Framework provides a `DeadlineManager` that allows working with deadlines. This is injected into the command handler method.

2. We use `deadlineManager` to schedule a named deadline (`"LC_APPROVAL_REMINDER"` in this case) that will expire in 10 days.

3. When the deadline is met, it will result in `LCApprovalPendingNotification`, which can be handled just like a command. Except, in this case, the behavior is triggered by the passage of time.

If no action is taken for 10 days, this is what we expect:

```
 1  class LCApplication {
 2
 3      @Test
 4      void shouldTriggerApprovalPendingEventTenDaysAfterSubmission() {
 5          final LCApplicationId id = LCApplicationId.randomId();
 6          fixture.given(new LCApplicationStartedEvent(id, ApplicantId.randomId(),
 7                      "My LC", LCState.DRAFT),
 8                  new LCAmountChangedEvent(id, THOUSAND_DOLLARS),
 9                  new MerchandiseChangedEvent(id, merchandise()))
10              .andGivenCommands(new SubmitLCApplicationCommand(id))  ❶
11              .whenThenTimeElapses(Duration.ofDays(10))              ❷
12              .expectDeadlinesMet(
13                  LCApprovalPendingNotification.first(id))          ❸
14              .expectEvents(new LCApprovalPendingEvent(id));        ❹
15      }
16  }
```

1. Given that the LC application is submitted.

2. When the period of 10 days elapses.

3. The deadline should be met.

4. `LCApprovalPendingEvent` should be emitted.

Let's look at how to implement this:

```
 1  import org.axonframework.deadline.annotation.DeadlineHandler;
 2
 3  class LCApplication {
 4
 5      @DeadlineHandler(deadlineName = "LC_APPROVAL_REMINDER")      ❶
 6      public void on(LCApprovalPendingNotification notification) {  ❷
 7
 8          AggregateLifecycle.apply(new LCApprovalPendingEvent(id));  ❸
 9
10      }
11  }
```

1. Deadlines are handled by annotating handler methods with the `@DeadlineHandler` annotation. Note that the same deadline name used previously is being referenced here.

2. This is the deadline handler method and uses the same payload that was passed along when it was scheduled.

3. We emit `LCApprovalPendingEvent` when the deadline expires.

The deadline handling logic should only be triggered if no action is taken. However, if the LC is either approved or rejected within a duration of 10 days, none of this behavior should be triggered:

```
 1  class LCApplicationAggregateTests {
 2      //...
 3      @Test
 4      void shouldNotTriggerPendingReminderIfApplicationIsApprovedWithinTenDays() {
 5          final LCApplicationId id = LCApplicationId.randomId();
 6          fixture.given(new LCApplicationStartedEvent(id, ApplicantId.randomId(),
 7                          "My LC", LCState.DRAFT),
 8                      new LCAmountChangedEvent(id, THOUSAND_DOLLARS),
 9                      new MerchandiseChangedEvent(id, merchandise()))
10                  .andGivenCommands(new SubmitLCApplicationCommand(id))  ❶
11
12                  .when(new ApproveLCApplicationCommand(id))            ❷
13                  .expectEvents(new LCApplicationApprovedEvent(id))
14                  .expectNoScheduledDeadlines();                       ❸
15      }
16
17      @Test
18      void shouldNotTriggerPendingReminderIfApplicationIsDeclinedWithinTenDays() {
19          // Test code is very similar. Excluded for brevity
20      }
21  }
22  }
```

1. Given that the LC application is submitted.

2. When it is approved within a duration of 10 days (in this case, almost immediately).

3. We expect no scheduled deadlines.

The implementation for this looks as follows:

```
 1  class LCApplication {
 2      //...
 3      @CommandHandler
 4      public void on(ApproveLCApplicationCommand command,
 5                      DeadlineManager deadlineManager) {
 6          assertInSubmitted(state);
 7          AggregateLifecycle.apply(new LCApplicationApprovedEvent(id));
 8          deadlineManager.cancelAllWithinScope("LC_APPROVAL_REMINDER"); ❶
 9      }
10
11      @CommandHandler
12      public void on(DeclineLCApplicationCommand command,
13                      DeadlineManager deadlineManager) {
14          assertInSubmitted(state);
15          AggregateLifecycle.apply(new LCApplicationDeclinedEvent(id));
16          deadlineManager.cancelAllWithinScope("LC_APPROVAL_REMINDER"); ❶
17      }
18
19      //...
20  }
```

1. We cancel all the deadlines with the name LC_APPROVAL_REMINDER (in this case, we only have one deadline with that name) within the scope of this aggregate.

Summary

In this chapter, we examined how to work with long-running workflows using sagas and the different styles we can use to implement them. We also looked at the implications of using explicit orchestration versus implicit choreography. We finally looked at how we can handle deadlines when there are no user-initiated actions.

You have learned how sagas can act as a first-class citizen in addition to aggregates when designing a system that makes use of domain-driven design principles.

In the next chapter, we will look at how we can interact with external systems while respecting bounded context boundaries between core and peripheral systems.

Further reading

Title	Author	Location
Saga persistence and event-driven architectures	Udi Dahan	`https://udidahan.com/2009/04/20/saga-persistence-and-event-driven-architectures/`
Sagas solve stupid transaction timeouts	Udi Dahan	`https://udidahan.com/2008/06/23/sagas-solve-stupid-transaction-timeouts/`
Microservices — when to react versus orchestrate	Andrew Bonham	`https://medium.com/capital-one-tech/microservices-when-to-react-vs-orchestrate-c6b18308a14c`
Saga orchestration for microservices using the outbox pattern	Gunnar Morling	`https://www.infoq.com/articles/saga-orchestration-outbox/`
Patterns for distributed transactions within a microservices architecture	Keyang Xiang	`https://developers.redhat.com/blog/2018/10/01/patterns-for-distributed-transactions-within-a-microservices-architecture`

9

Integrating with External Systems

Wholeness is not achieved by cutting off a portion of one's being, but by integration of the contraries.

– Carl Jung

So far, we have used DDD to implement a robust core for our application. However, most solutions (by extension-bounded contexts) usually have both upstream and downstream dependencies that usually change at a pace, which is different from these core components. To maintain both agility and reliability and enable loose coupling, it is important to integrate with a peripheral system in a manner that shields the core from everything else that surrounds it.

In this chapter, we will look at the LC application processing solution and examine the means by which we can integrate with other components in the ecosystem. You will learn how to recognize relationship patterns between components.

This chapter covers the following main topics:

- Continuing our design journey
- Bounded context relationships
- Implementation patterns

By the end of the chapter, we will round off by looking at common patterns when integrating with legacy applications. Let's dive right in!

Continuing our design journey

From our domain analysis in earlier chapters, we have arrived at four bounded contexts for our application, as depicted here:

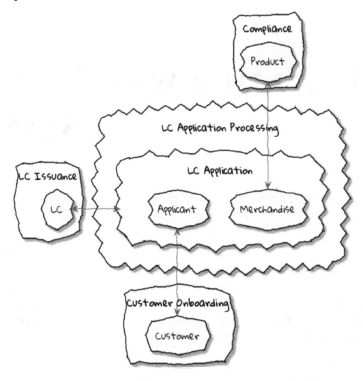

Figure 9.1 – Relationship between bounded contexts

So far, our focus has been on the implementation of the internals of the **LC application** bounded context. While the LC application bounded context is independent of the other bounded contexts, it is not completely isolated from them. For example, when processing an LC application, we need to perform merchandise and applicant checks that require interactions with the **compliance** and **customer onboarding** bounded contexts respectively. This means that these bounded contexts have a relationship with each other. These relationships are driven by the nature of collaboration between the teams working on the respective bounded contexts. Let's examine how these team dynamics influence integration mechanisms between bounded contexts in a way that continues to preserve their individual integrity.

Bounded context relationships

We need bounded contexts to be as independent as possible. However, this does not mean that bounded contexts are completely isolated from each other. Bounded contexts need to collaborate with others to provide business value. Whenever there is collaboration required between two bounded contexts, the nature of their relationship is not only influenced by their individual goals and priorities but also by the prevailing organizational realities. In a high-performing environment, it is fairly common to have a single team assume ownership of a bounded context. The relationships between the teams owning these bounded contexts play a significant role in influencing the integration patterns employed to arrive at a solution. At a high level, there are two categories of relationships:

- Symmetric
- Asymmetric

Let's look at these relationship types in more detail.

Symmetric relationship patterns

Two teams, say team A and team B, can be said to have a symmetric relationship when they have an equal amount of influence in the decision-making process to arrive at a solution. Both teams are in a position to, and, indeed, do, contribute more or less equally toward the outcome. Here's a diagrammatic representation:

Figure 9.2 – Both teams have an equal say in influencing the solution

There are three variations of symmetric relationships, each of which we will outline in more detail in the following subsections.

Partnership

In a partnership, both teams integrate in an ad hoc manner. There are no fixed responsibilities assigned when needing complete integration work. Each team picks up work as and when needed without the need for any specific ceremony or fanfare. The nature of the integration is usually two-way, with both teams exchanging solution artifacts as and when needed. Such relationships require extremely high degrees of collaboration and understanding of the work done by both teams. Check the following figure:

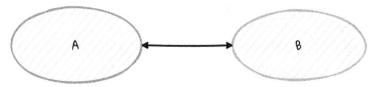

Figure 9.3 – There is an ad hoc mutual dependency between teams in a partnership relationship

Example

Let's take as an example a web frontend team working in close collaboration with an API team building the BFFs (https://philcalcado.com/2015/09/18/the_back_end_for_front_end_pattern_bff.html) for the frontend. The BFF team creates experience APIs meant to be used exclusively by the frontend. To fulfill any functionality, the frontend team requires capabilities to be exposed by the APIs team. On the other hand, the APIs team is dependent on the frontend team to provide advice on what capabilities to build and the order in which to build them. Both teams freely make use of each other's domain models (for example, the same set of request and response objects that define the API) to implement functionality. Such reuse happens mostly arbitrarily, and when API changes happen, both teams coordinate changes to keep things working.

When to use

Partnership between teams requires high levels of collaboration, trust, and understanding. Teams tend to use this partnership when team boundaries are informal. It also helps if these teams are co-located and/or have a significant working time overlap.

Potential pitfalls

Partnership relationships between teams can lead to a situation where individual team responsibilities become very unclear, leading a solution toward the dreaded *big ball of mud*.

Shared kernel

Unlike in a partnership, when using a shared kernel, teams have a clear understanding of the solution artifacts and models they choose to share between themselves. Both teams take equal responsibility for the upkeep of these shared artifacts.

Example

The *LC application processing* and *customer onboarding* teams in our LC application may choose to use a common model to represent `CustomerCreditValidatedEvent`. Any enhancements or changes to the event schema can affect both teams. The responsibility to make any changes is owned by both teams. Intentionally, these teams do not share anything beyond these mutually agreed-upon models and artifacts. Here's a representation of a shared kernel relationship between teams:

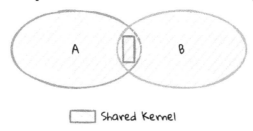

Figure 9.4 – Teams have an explicit understanding of shared models

When to use

The shared kernel form of collaboration works well if shared artifacts are required to be consumed in an identical fashion in both contexts. Furthermore, it is attractive for multiple teams to coordinate and continue sharing, as opposed to duplicating identical models in both contexts.

Potential pitfalls

Changes made to the shared kernel affect both bounded contexts. This means that any change made to the shared kernel needs to remain compatible for both teams. Needless to say, as the number of teams using the shared kernel increases, the cost of coordination goes up manifold.

Separate ways

When two teams choose to not share any artifacts or models between them, they go their own separate ways:

Figure 9.5 – Teams go separate ways and don't share anything between them

Example

The *LC application processing* and *customer onboarding* teams may start by sharing the same build/deployment scripts for their services. Over a period of time, deployment requirements may diverge to a point where the shared cost of maintaining these scripts becomes prohibitively expensive, causing these teams to fork their deployments to regain independence from the other team.

When to use

In some cases, two teams may be unable to collaborate for a variety of reasons, ranging from a drift in individual team requirements to organizational politics. Whatever the case may be, these teams may decide that the cost of collaboration is too high, resulting in them going their own separate ways.

Potential pitfalls

Choosing to go separate ways may result in duplicate work across affected bounded contexts. When working in bounded contexts that map to core subdomains, this may prove counterproductive, as it can lead to inconsistent behaviors unintentionally.

It is possible to transition from one relationship type to another over a period of time. In our experience, transitioning from any one of these relationships may not be straightforward. In cases where requirements are relatively clear at the outset, it may be easier to start with a *shared kernel*. Conversely, if requirements are unclear, it may be prudent to start either with a loose *partnership* or go *separate ways* until requirements become clear. In any of these scenarios, it is important to keep evaluating the nature of the relationship and transition to a more appropriate type, based on our enhanced understanding of the requirements and/or the relationship itself.

In each of the preceding characterized relationships, the teams involved have a more or less equal say in how the relationship evolved and the resulting outcomes. However, this may not always be the case. Let's look at examples of cases where one team may have a clear upper hand in terms of how the relationship evolves.

Asymmetric relationship patterns

Two teams can be said to have an asymmetric relationship when one of the teams has a stronger influence in the decision-making process to arrive at a solution. In other words, there is a clear customer-supplier (or upstream-downstream) relationship where either the customer or the supplier plays a dominant role that affects solution design approaches. It is also likely that the customer and the supplier do not share common goals. Here is a representation of an asymmetric relationship between customer and supplier:

Figure 9.6 – One of the teams has a dominant say in influencing the solution

There are at least three solution patterns when teams are in an asymmetric relationship, each of which we will outline in more detail in the following subsections.

Conformist (CF)

It is not unusual for the side playing the supplier role to have a dominant say in how the relationship with one or more customers is implemented. Furthermore, the customer may simply choose to conform to the supplier-provided solution as is, making it an integral part of their own solution. In other words, the supplier provides a set of models and the customer uses those same models to build their solution. In this case, the customer is considered to be a *conformist*:

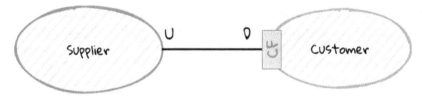

Figure 9.7 – The customer accepts dependency on the supplier model

Example

When building a solution to validate the United States postal addresses of LC applicants, we chose to conform to the USPS Web Tools address validation API schema (https://www.usps.com/business/web-tools-apis/). Given that the business started with just US-based applicants, this made sense. This means that any references to the address model in our bounded contexts mimic the schema prescribed by the USPS. Furthermore, it means that we will need to keep up with changes that occur in the USPS API as and when they occur (regardless of whether those changes are needed for our own functionality).

When to use

Being a conformist is not necessarily a negative thing. The supplier's models may be a well-accepted industry standard, or they may simply be good enough for our needs. It may also be that the team may not have the necessary skills, motivation, or immediate need to do something different from what the supplier has provided. This approach also enables teams to make quick progress, leveraging work mostly done by other experts.

Potential pitfalls

An overuse of the conformist pattern may dilute the ubiquitous language of our own bounded contexts, resulting in a situation where there is no clear separation between the supplier and customer concepts. It may also be that concepts that are core to the supplier's context leak into our own, despite those concepts carrying little to no meaning in our context. This may result in these bounded contexts being very tightly coupled with each other. And if a need arises to switch to another supplier or support multiple suppliers, the cost of change may be prohibitively expensive.

Anti-Corruption Layer

There may be scenarios where a customer may need to collaborate with a supplier but may want to shield themselves from the supplier's ubiquitous language and models. In such cases, it may be prudent to redefine these conflicting models in the customer's own ubiquitous language using a translation layer at the time of integration, also known as an **Anti-Corruption Layer** (ACL). See the following figure:

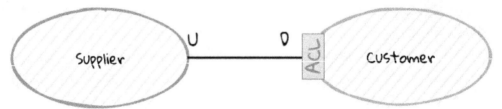

Figure 9.8 – The customer wants to protect themselves from supplier models

Example

In the address validation example referenced in the *Conformist (CF)* section, the *LC application processing* team may need to support Canadian applicants as well. In such a case, being a conformist to a system that supports only US addresses may prove restrictive and even confusing. For example, the US *state* is analogous to a *province* in Canada. Similarly, a *ZIP code* in the US is called a *postal code* in Canada. In addition, US ZIP codes are numeric whereas Canadian postal codes are alphanumeric. Most importantly, we currently do not have the notion of a *country code* in our address model, but now we will need to introduce this concept to differentiate addresses within the respective countries. Let's look at the address models from the respective countries here:

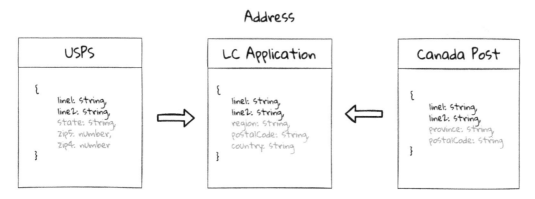

Figure 9.9 – The address models of different countries

While we initially conformed to the USPS model, we have now evolved to support more countries. For example, *region* is used to represent the concept of *state/province*. Also, we have introduced the *country* value object, which was missing earlier.

When to use

ACLs come in handy when the customer models are part of a core domain. The ACL shields the customer from changes in the supplier's models and can help produce more loosely coupled integrations. It may also be necessary when we are looking to integrate similar concepts from multiple suppliers.

Potential pitfalls

Using an ACL may be tempting in a lot of cases. However, it is less beneficial when the concepts being integrated don't often change or are defined by a well-known authority. Using an ACL with a custom language may only cause more confusion. Creating an ACL usually requires additional translations and thereby may increase the overall complexity of the customer's bounded context and may be considered premature optimization.

Open host service

Unlike the conformist and the ACL, where customers do not have a formal means to interface with the supplier, with the **Open Host Service (OHS)**, the supplier defines a clear interface to interact with its customers. This interface may be made available in the form of a well-known published language (for example, a REST interface or a client SDK):

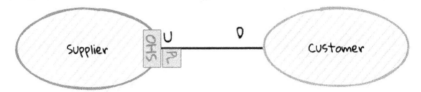

Figure 9.10 – An OHS using a Published Language (PL)

Example

The LC application processing bounded context can expose an HTTP interface for each of its commands, as shown here:

```
1   # Start a new LC application
2   curl POST /applications/start \
3           -d '{"applicant-id": "8ed2d2fe", "clientReference": "Test LC"}' \
4           -H 'content-type:application/vnd.lc-application.v2+json'
5
6   # Change the amount on an existing application
7   curl POST /applications/ac130002/change-amount \
8           -d '{"amount": 100, "currency": "USD"}' \
9           -H 'content-type:application/vnd.lc-application.v2+json'
10
11  # Other commands omitted for brevity
```

As an addition to the HTTP interface shown here, we can even provide a client SDK in some of the more popular languages used by our customers. This helps hide more implementation details such as the MIME type and version from customers.

When to use

When the supplier wants to hide its internal models (ubiquitous language), making an OHS enables the supplier to evolve while providing a stable interface to its customers. In a sense, the OHS pattern is a reversal of the ACL pattern – instead of the customer, the supplier implements the translation of its internal model. Also, the supplier can consider providing an OHS when it is interested in providing a richer user experience for its customers.

Potential pitfalls

While suppliers may have good intentions by providing an OHS for its customers, it may result in increased implementation complexity (for example, there may be a need to support multiple versions of an API, or client SDKs in multiple languages). If the OHS does not take into account the common usage patterns of its customers, it may result in poor customer usability and also in degraded performance for the supplier.

It is important to note that the conformist and the ACL are patterns that customers implement, whereas the OHS is a supplier-side pattern. For example, the following scenario where the supplier provides an *OHS* to one customer who is a *conformist* and another who has an *ACL* can be true, as depicted here:

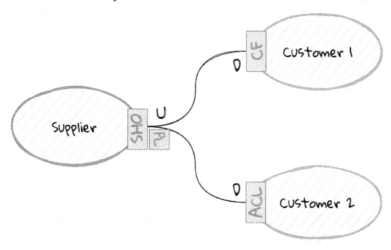

Figure 9.11 – Asymmetric relationships with multiple customers

Now that we have seen the various ways in which bounded contexts can integrate with each other, here is one possible implementation for our LC application, depicted in the form of a context map:

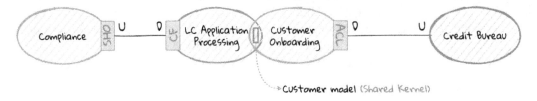

Figure 9.12 – A simplified context map for the LC application

So far, we have examined the various ways in which inter-team dynamics influence integration mechanisms. While having clarity at the conceptual level helps, let's see how these relationships manifest themselves at the implementation level.

Implementation patterns

We have looked at integration between bounded contexts at a design level, but these concepts need to be translated into code. There are three broad categories that can be employed when integrating two bounded contexts:

- Data-based
- Code-based
- API-based

Let's look at each method in more detail now.

Data-based

In this style of integration, the bounded contexts in question share data with each other. If the relationship is symmetric, the teams owning these bounded contexts may choose to share entire databases with free access to read, write, and change underlying structures. Conversely, in an asymmetric relationship, the supplier may constrain the scope of access, based on the type of relationship.

Shared database

The simplest form of data integration is the use of a shared database. In this style of integration, all participating bounded contexts have unrestricted access to the schemas and the underlying data, as shown here:

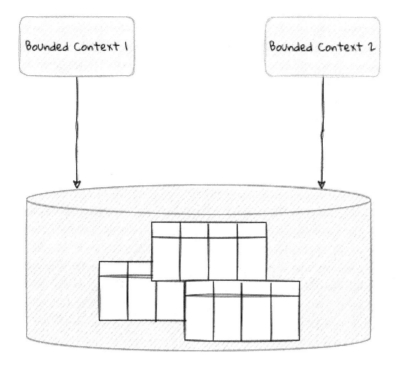

Figure 9.13 – Integration using a shared database

When to use

The shared database presents a very low barrier to entry for teams looking to quickly enable new or enhance existing functionality by providing ready access to data for read and/or write use cases. More importantly, it also allows the use of local database transactions, which usually provides strong consistency, lower complexity, and better performance (especially when working with relational databases).

Potential pitfalls

However, this symmetric integration style where multiple teams have shared ownership is usually frowned upon because it often leads to a situation where there is no clear ownership. Furthermore, the shared databases can become a source of tight coupling, accelerating the path toward the dreaded *big ball of mud*. Additionally, users of the shared database can suffer from the *noisy neighbor* effect where one co-tenant monopolizing resources adversely affects all other tenants. For these reasons, teams will be well advised to choose this style of integration sparingly.

Replicated data

In the case of asymmetric relationships, suppliers may be unwilling to provide direct access to their data. However, they may choose to integrate with customers using a mechanism based on data sharing. An alternate form of integration is to provide a copy of the data required by consumers. There are many variations on how this can be implemented; we depict the more common ways here:

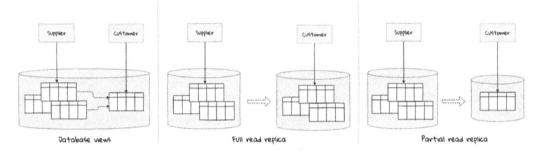

Figure 9.14 – Integration using data replication

- **Database views**: In this form, the consumer gets or is provided access to a subset of data using query-based or materialized views. In either case, the customer usually has read-only access to the data, and both supplier and customer continue to share the same physical resources (usually the database engine).

- **Full read replica**: In this form, the customer gets access to a read replica of the supplier's entire database, usually on a physically disparate infrastructure.

- **Partial read replica**: In this form, the customer gets access to a read replica of a subset of the supplier's database, again on a physically disparate infrastructure.

When to use

This style of integration may be required when there is an asymmetric relationship between the supplier and the customer. Like the shared database, this integration style usually requires less upfront effort to integrate. This is also apt when suppliers intend to provide read-only access to a subset of their data. It may also suffice to use data replication when customers are only required to read a subset of the supplier's data.

Potential pitfalls

If we choose to use database views, we may continue to suffer from the noisy neighbor effect. On the other hand, if we choose to create physically disparate replicas, we will need to incur the cost of additional operational complexity. More importantly, the consumers remain tightly coupled to the supplier's domain models and ubiquitous language.

Next, let's look at some ways to make the most of data-based integrations.

Increasing effectiveness

When sharing data, the schema (the structure of the database) acts as a means to enforce contracts, especially when using databases that require specifying a formal structure (for example, relational databases). When multiple parties are involved, managing the schema can become a challenge.

To mitigate undesirable changes, teams sharing data may want to consider the use of a schema migration tool. Relational databases work well with tools such as Liquibase (`https://www.liquibase.org/`) or Flyway (`https://flywaydb.org/`). When working with databases that do not formally enforce a schema, it may be best to avoid employing this style of integration, especially when working in symmetric relationships where ownership is unclear.

In any case, if using one of the shared data styles of integration is unavoidable, teams may want to strongly consider employing one or more of the aforementioned techniques in refactoring databases to make it more manageable.

Code-based

In this style of integration, teams coordinate by sharing code artifacts, either directly in the form of source code and/or binaries. At a high level, there are two forms:

- Sharing source code
- Sharing binaries

We will describe each of these here.

Sharing source code

A fairly common practice within organizations is to share source code with the objective of promoting reuse and standardization. This may include utilities (such as logging and authentication), build/deployment scripts, and data transfer objects – in other words, any piece of source code where the cost of duplication is seen to be higher than reuse.

When to use

Depending on the relationship type (symmetric/asymmetric), teams sharing code may have varying levels of influence on how the shared artifacts evolve. This works well in a symmetric relationship, where both teams are empowered to make changes compatible with each other. Similarly, in an asymmetric relationship, the supplier may accept changes from customers, while retaining ownership and control of the shared artifacts. This also tends to work well in the case of non-core, infrequently changing code artifacts. Sharing source code also enables higher levels of transparency and visibility for the internals of the shared artifacts (a case in point is open source software).

Potential pitfalls

Sharing code artifacts means that individual teams take on the responsibility to make sure that the process of converting source code into binary executables is uniform and compatible with the requirements of all parties. This may include code conventions, static quality checks, tests (the presence or lack thereof), compilation/build flags, and versioning. When a relatively large number of teams is involved, maintaining this form of compatibility may become burdensome.

Sharing binary artifacts

Another relatively common practice is to share artifacts at the binary level. In this scenario, the consumers may or may not have direct access to source code artifacts. Examples include third-party libraries, client SDKs, and API documentation. This form of integration is fairly common when the relationship between the coordinating parties is asymmetric. The supplier of the library has clear ownership of maintaining the life cycle of the shared artifacts.

When to use

Sharing just binary artifacts may be necessary when the supplier is unable/unwilling to share source artifacts, possibly because they may be proprietary and/or part of the supplier's intellectual property. Because the supplier takes ownership of the *build* process, it behooves the supplier to produce artifacts that are compatible with most potential consumers. Hence, this works well when the supplier is willing to do that. On the other hand, it means that the customer places high levels of trust (https://www.thoughtworks.com/en-us/insights/podcasts/technology-podcasts/securing-software-supply-chain) in the supplier's software supply chain (https://blog.sonatype.com/software-supply-chain-a-definition-and-introductory-guide) when producing these artifacts.

Potential pitfalls

Integration through the use of binary artifact sharing reduces the visibility of the build process of the shared artifacts for consumers. If consumers rely on slow-moving suppliers, this can become untenable. For example, if a critical security bug is discovered in the shared binary, the consumer is solely reliant on the supplier to remediate it. This can be a huge risk if such dependencies are in critical, business-differentiating aspects of the solution (especially in the core subdomain). This risk can be exacerbated without the use of appropriate ACLs and/or **Service-Level Agreements** (**SLAs**).

Increasing effectiveness

When sharing code artifacts, it becomes a lot more important to be explicit in how changes are made while continuing to maintain high levels of quality – especially when multiple teams are involved. Let's examine some of these techniques in more detail:

- **Static analysis**: This can be as simple as adhering to a set of coding standards using a tool such as Checkstyle. More importantly, these tools can be used to conform to a set of naming conventions to allow the firmer use of the ubiquitous language throughout the code base. In addition, tools such as SpotBugs and PMD/CPD can be used to statically analyze code for the presence of bugs and duplicate code.

- **Code architecture tests**: While static inspection tools are effective at operating at the level of a single compilation unit, runtime inspection can take this one level further to identify package cycles, dependency checks, inheritance trees, and so on to apply lightweight architecture governance. The use of tools such as JDepend and ArchUnit can help here.

- **Unit tests**: When working with shared code bases, team members are looking to make changes safely and reliably. The presence of a comprehensive suite of fast-running unit tests can go a long way toward increasing confidence. We strongly recommend employing test-driven design to further maximize creating a code base that is well designed and enables easier refactoring.

- **Code reviews**: While automation can go a long way, augmenting the process where a human reviews changes can be highly effective for multiple reasons. This can take the form of offline reviews (using pull requests) or active peer reviews (using paired programming). All of these techniques serve to enhance collective understanding, thereby reducing risk when changes are made.

- **Documentation**: Needless to say, well-structured documentation can be invaluable when making contributions and also when consuming binary code artifacts. Teams will be well advised to proliferate the use of the ubiquitous language by striving to write self-documenting code throughout to maximize the derived benefits.

- **Dependency management**: When sharing binary code artifacts, managing dependencies can become fairly complicated due to having too many dependencies, long dependency chains, conflicting/cyclic dependencies, and so on. Teams should strive to reduce afferent (incoming) coupling as much as possible to mitigate the problems described previously.

- **Versioning**: In addition to minimizing the amount of afferent coupling, using an explicit versioning strategy can go a long way toward making dependency management easier. We strongly recommend considering the use of a technique such as semantic versioning for shared code artifacts.

IPC-based

In this style of integration, the bounded contexts exchange messages using some form of **Inter-Process Communication** (**IPC**) to interact with each other. This can take the form of synchronous or asynchronous communication.

Synchronous messaging

Synchronous messaging is a style of communication where the sender of the request waits for a response from the receiver, which implies that the sender and the receiver need to be active for this style to work. Usually, this form of communication is point to point. HTTP is one of the commonly used protocols for this style of communication. A visual representation of this form of communication is shown here:

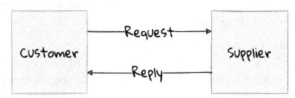

Figure 9.15 – Synchronous messaging

> **Note**
> Please take a look at the HTTP APIs for the commands used during LC application processing, which are included with the code examples for this chapter.

When to use

This form of integration is used when the customer is interested in the supplier's response to the request. The response is then used to determine whether the request was successful or not. Given that the customer needs to wait for the response, it is advisable to use this style of messaging for low-latency operations. This form of integration is popular when exposing public APIs over the internet (for example, GitHub's REST API, which you can learn more about at https://docs.github.com/en/rest).

Potential pitfalls

When using synchronous messaging, the customer's ability to scale is heavily dependent on the supplier to satisfy the customer's requirements. On the flip side, customers making requests at too high a rate may compromise the supplier's ability to serve customers in a predictable manner. If there is a chain of synchronous messaging, the probability of cascading failure becomes much higher.

Asynchronous messaging

Asynchronous messaging is a style of communication where the sender does not wait for an explicit response from the receiver.

> **Note**
>
> We are using the terms *sender* and *receiver* instead of *customer* and *supplier* because they both can play the role of sender or receiver.

This is typically achieved by introducing an intermediary in the form of a message channel. The presence of the intermediary enables both one-to-one and one-to-many modes of communication. Typically, the intermediary can take the form of a shared filesystem, database, or queueing system:

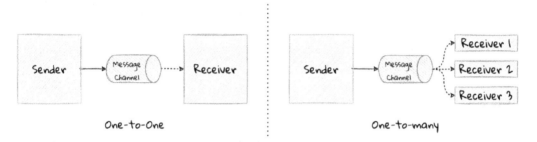

Figure 9.16 – Asynchronous messaging

> **Note**
>
> Please take a look at the event APIs for the commands used during LC application processing, which are included with the code examples for this chapter.

When to use

This form of integration is used when the sender does not care about receiving an **immediate** response from the receiver(s), resulting in the respective systems becoming a lot more decoupled from each other. This further enables these systems to scale independently. This also makes it possible to have the same message being processed by multiple receivers. For example, in our LC application processing system, LCApplicationSubmittedEvent is received by both the *compliance* and *customer onboarding* systems.

Potential pitfalls

The introduction of the intermediary component adds complexity to the overall solution. The non-functional characteristics of the intermediary can have a profound effect on the resilience characteristics of the system as a whole. It can also be tempting to add processing logic to the intermediary, thereby coupling the overall system very tightly to this component. To ensure reliable communication between the sender and the receiver, the intermediary may have to support a variety of enhanced capabilities (such as ordering, producer flow control, durability, and transactions).

Increasing effectiveness

When implementing integration using some form of IPC, a lot of the techniques discussed in the code-based implementation patterns section continue to apply. As discussed earlier, API documentation plays a significant role in reducing friction for customers. In addition, here are a few more techniques that apply specifically when using IPC-based integration:

- **Typed protocols**: When working with this form of integration, it is important to minimize the amount of time taken to gather feedback on structural validations. This is especially critical given that the supplier and the customer may be in a constant state of independent evolution. The use of typed protocols such as Protocol Buffers, Avro, Netflix's Falcor, and GraphQL can make it easier for customers to interact with suppliers while maintaining a lightweight mechanism to validate whether requests are correct.

> **Note**
>
> The key word here is **lightweight**. It is pertinent to note that we are not advising against the use of JSON-based HTTP APIs (typically advertised as being RESTful), which do not enforce the use of an explicit schema. Neither are we promoting the use of (arguably) legacy protocols such as SOAP, WSDL, and CORBA. Each of these, while being well-meaning, suffered from being fairly heavyweight.

- **Self-discovery**: As outlined previously, when working with an IPC-based integration mechanism, we should look to reduce the barrier to entry. When working with RESTful APIs, the use of HATEOAS (`https://restfulapi.net/hateoas`), although difficult for suppliers to implement, can make it easier for customers to understand and consume APIs. In addition, making use of a service registry and/or a schema registry can further reduce consumption friction.

- **Contract tests**: In the spirit of failing fast and shifting left, the practice of contract testing and consumer-driven contracts can further increase the quality and speed of integration. Tools such as Pact (`https://pact.io/`) and Spring Cloud Contract (`https://spring.io/projects/spring-cloud-contract`) make the adoption of these practices relatively simple.

So far, we've discussed implementation patterns, broadly categorized into data-based, code-based, and IPC-based integrations. Hopefully, this gives you a good start in consciously choosing the appropriate approach by considering the benefits and caveats that they bring along with them.

Summary

In this chapter, we looked at the different types of bounded context relationships. We also examined common integration patterns that can be used when implementing these bounded context relationships.

You have learned when specific techniques can be used, about potential pitfalls, and ideas on how to increase effectiveness when employing these methods.

In the next chapter, we will explore the means to distribute these bounded contexts into independently deployable components (in other words, employ a microservices-based architecture).

Further reading

Title	Author	Location
Integration database	Martin Fowler	`https://martinfowler.com/bliki/IntegrationDatabase.html`
REST APIs must be hypertext-driven	Roy T. Fielding	`https://roy.gbiv.com/untangled/2008/rest-apis-must-be-hypertext-driven`

Part 3:
Evolution Patterns

In the previous section, we built an application from scratch. However, we packaged all components in a single deployable as a monolith. In this part, we will extend the application we built in *Part 2* by exploring various options on how we can decompose this application iteratively into finer-grained components. We will also look at the implications of decomposition along both functional and cross-functional aspects.

This part contains the following chapters:

- *Chapter 10, Beginning the Decomposition Journey*
- *Chapter 11, Decomposing into Finer-Grained Components*
- *Chapter 12, Beyond Functional Requirements*

10
Beginning the Decomposition Journey

A distributed system is one in which the failure of a computer you didn't even know existed can render your own computer unusable.

— *Leslie Lamport*

So far, we have a working application for **Letter of Credit** (**LC**) application processing, which is bundled along with other components as a single package. Although we have discussed the idea of subdomains and bounded contexts, the separation between these components is logical rather than physical. Furthermore, we have primarily focused on the *LC Application Processing* aspect of the overall solution.

In this chapter, we will look at how to extract the LC Application Processing bounded context into a component that is physically disparate and, hence, enables us to deploy them independently of the rest of the solution. We will discuss the various options that are available to us, the rationale for choosing a given option, and the implications that we need to be cognizant of.

In this chapter, we'll cover the following topics:

- Continuing our design journey
- Decomposing our monolith
- Changes to frontend interactions
- Changes in database interactions

By the end of this chapter, you will have learned what it takes to design well-factored APIs—both remote procedure calls and event-based. For event-based APIs, you will gain an understanding of the various guarantees that might be needed to create robust solutions. Finally, you will also learn how to manage consistency when using multiple data stores.

Continuing our design journey

In the preceding chapters, we had a solution for LC Application Processing that worked as an in-process component of the remainder of the overall application. From a logical perspective, our realization of the LC application is similar to the following diagram:

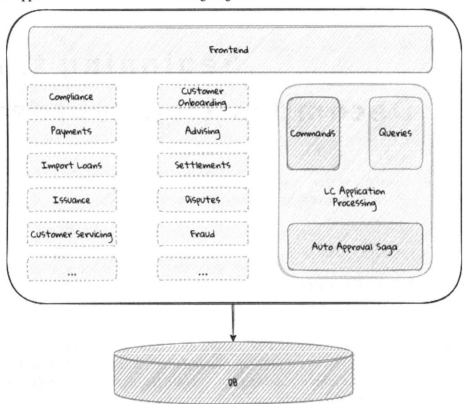

Figure 10.1 – The current view of the LC application monolith

Although the **LC Application Processing** component is loosely coupled with the rest of the application, we are still required to coordinate with several other teams to realize the business value. This could inhibit our ability to innovate at a pace that is faster than the slowest contributor in the ecosystem. This is because all teams need to be production-ready before a deployment can happen. This can be further exacerbated by the fact that individual teams might be at different levels of engineering maturity. Let's look at some options regarding how we can achieve a level of independence from the rest of the ecosystem by physically decomposing our components into distinctly deployable artifacts.

Decomposing our monolith

First and foremost, the **LC Application Processing** component exposes only in-process APIs when other components interact with it. This includes interactions with the following:

- The frontend

- Published/consumed events

- Databases

To extract LC Application Processing functionality into its own, independently deployable component, remotely invokable interfaces will have to be supported instead of the in-process ones that we currently have. So, let's examine the remote API options for each.

Changes to frontend interactions

Currently, the **JavaFX** frontend interacts with the rest of the application by making request-response style in-process method calls (that is, `CommandGateway` for commands and `QueryGateway` for queries), as shown here:

```
 1  @Service
 2  public class BackendService {
 3
 4      private final QueryGateway queryGateway;
 5      private final CommandGateway commandGateway;
 6
 7      public BackendService(QueryGateway queryGateway,
 8                            CommandGateway commandGateway) {
 9          this.queryGateway = queryGateway;
10          this.commandGateway = commandGateway;
11      }
12
13      public LCApplicationId startNewLC(ApplicantId applicantId,
14                                        String clientReference) {
15          return commandGateway.sendAndWait(
16                  startApplication(applicantId, clientReference));
17      }
18
19      public List<LCView> findMyDraftLCs(ApplicantId applicantId) {
20          return queryGateway.query(
21                  new MyDraftLCsQuery(applicantId),
22                      ResponseTypes.multipleInstancesOf(LCView.class))
23                  .join();
24
25      }
26  }
```

One very simple way to replace these in-process calls is to introduce some form of **Remote Procedure Call** (**RPC**). Now our application looks similar to the following:

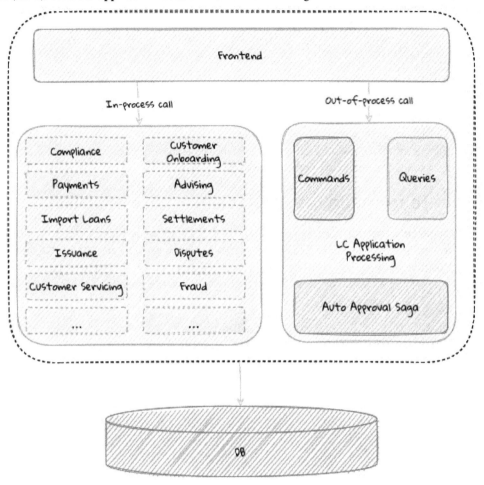

Figure 10.2 – Introducing remote interaction with the frontend

When working with in-process interactions, we are simply invoking methods on objects within the confines of the same process. However, when we switch to using out-of-process calls, there are quite a few considerations. These days when working with remote APIs, we have several popular choices in the form of **JSON**-based web services, **GraphQL**, **gRPC**, and more. While it is possible to make use of a completely custom format to facilitate the communication, DDD advocates the use of the **Open Host Service pattern** (https://ddd-book.karthiks.in/10-distributing-into-multiple-components.html#_open_host_service_ohs) using a published language that we covered in *Chapter 9, Integrating with External Systems*. Even with the open host service style of communication, there are a few considerations, some of which we discuss in the following subsections.

Protocol options

There are several options available to us when exposing remote APIs. These days, using a JSON-based API (often labeled as **Representation State Transfer** or **REST**) seems to be quite popular. However, this isn't the only option available to us. In a resource-based approach, the first step is to identify a resource (noun) and then map the interactions (verbs) associated with the resource as a next step. In an action-based approach, the focus is on the actions to be performed. Arguably, REST takes a resource-based approach, whereas **graphQL**, **gRPC**, **SOAP**, and more, seem to be action-based. Let's take an example of an API where we want to start a new LC application. In a RESTful world, this could look something like this:

```
1   # Start a new LC application
2   curl POST /lc-applications/start-new \
3           -d '{"applicant-id": "8ed2d2fe", \
4               "clientReference": "Test LC"}' \
5           -H 'content-type:application/vnd.lc-application.v2+json'
```

In comparison, with a graphQL implementation, this could look like the following:

```
1   mutation StartLCApplication {
2     startLCApplication(applicantId: "8ed2d2fe",
3                        clientReference: "Test LC") {
4       lcApplicationId
5     }
6   }
```

In our experience, designing APIs using REST does result in some form of dilution when attempting to mirror the language of the domain—because the focus is first and foremost on resources. Purists will be quick to point out that the preceding example is not RESTful because there is no resource named `start-new` and that we should leave the URL to simply include the name of the resource (use /lc- applications instead of `/lc-applications/start-new`). Our approach is to place more importance on remaining true to the ubiquitous language as opposed to being dogmatic about adherence to technical purity.

Transport format

Here, we have two broad choices: **text-based** (for example, **JSON** or **XML**) versus **binary** (for example, protocol buffers, `https://developers.google.com/protocol-buffers`, or Avro, `https://avro.apache.org/`). If non-functional requirements (such as performance) are met, our preference is to use text-based protocols as a starting point. That's because it can afford the flexibility of not needing any additional tools to visually interpret the data (when debugging).

When designing a remote API, we have the option of choosing a format that enforces a schema (for example, protocol buffers or Avro) or something less formal such as plain JSON. In such cases, in order to stay true to the ubiquitous language, the process might have to include additional governance in the form of more formal design and code reviews, documentation, and more.

Compatibility and versioning

As requirements evolve, there will be a need to enhance the interfaces to reflect these changes. This will mean that our ubiquitous language will also change over time, rendering old concepts obsolete. The general principle is to maintain backward compatibility with consumers for as long as possible. But this does come at the cost of having to maintain old and new concepts together—leading to a situation where it can become hard to tell what is relevant versus what is not. Using an explicit versioning strategy can help to manage this complexity up to an extent—where newer versions might be able to break backward compatibility with older ones. However, it is also not feasible to continue supporting a large number of incompatible versions indefinitely. Hence, it is important to make sure that the versioning strategy makes deprecation and retirement agreements explicit.

REST APIs

We recognize that there are several options when exposing web-based APIs, and claims of using a REST approach seem quite common these days. REST was coined by Roy Fielding as part of his doctoral dissertation. The idea of what constitutes REST has been a matter of debate and, arguably, remains ambiguous even today. Leonard Richardson introduced the notion of a maturity model for HTTP-based REST APIs that somewhat helped provide some clarity. The model describes broad conformance to REST in four levels, with each level being more mature than the preceding one:

0.   **Adhoc**: Where APIs are designed without the use of any perceptible structure.

1.   **Resources**: Where APIs are designed around a *thing* that makes sense on its own (usually, this is a noun). Here, a very small subset of verbs (either a GET or a POST) could be used to model all operations.

2.   **HTTP verbs**: Where APIs are designed by making use of a standard set of operations that can be performed on a resource (for example, GET for reads, POST for creates, PUT for updates, DELETE for deletes, and more).

3.   **HATEOAS**: Where APIs include hypermedia links to help clients discover our API in a self-service manner.

In our experience, most web service-based solutions that claim to be RESTful seem to stop at level 2. Roy Fielding, the inventor of REST, seems to claim that *REST APIs must be hypertext-driven* (`https://roy.gbiv.com/untangled/2008/rest-apis-must-be-hypertext-driven`). In our opinion, the use of hypertext controls in APIs allows them to become self-documenting and, thereby, promotes the use of the ubiquitous language more explicitly. More importantly, it also indicates what operations are applicable for a given resource at that time in its life cycle. For example, let's look at a sample response where all pending LC applications have been listed:

```
1   GET /lc-applications?status=pending HTTP/1.1
2   Content-Type: application/json
3
4   HTTP/1.1 200 OK
5   Content-Type: application/prs.hal-forms+json
6   {
7     "_embedded" : {
8       "lc-applications" : [
9         {
10         "clientReference" : "Test LC",
11         "_links" : {
12           "self" : {
13             "href" : "/lc-applications/582fe5f8"
14           },
15           "submit" : {
16             "href" : "/lc-applications/582fe5f8/submit"
17           }
18         }
19       },
20       {
21         "clientReference" : "Another LC",
22         "_links" : {
23           "self" : {
24             "href" : "/lc-applications/7689da3e"
25           },
26           "approve" : {
27             "href" : "/lc-applications/7689da3e/approve"
28           },
29           "reject" : {
30             "href" : "/lc-applications/7689da3e/reject"
31           }
32         }
33       }
34     ]
35   }
36 }
```

In the preceding example, there are two `lc-applications` listed. Based on the current status of the LC, the links provide a means to act on the LC appropriately. In addition to the *self* link, the first LC application shows a submit link denoting that it can be submitted, whereas the second application shows the approve and reject links, but not a submit link. Presumably, this is because it has already been submitted. Also, notice how the response does not need to include a status attribute so that they can use this to deduce which operations are relevant for the LC application at that time (this is an example of the *tell, don't ask* principle, `https://martinfowler.com/bliki/TellDontAsk.html`). While this might be a subtle nuance, we felt that it is valuable to point out in the context of our DDD journey.

So, we have looked at a few considerations when moving from an in-process API to an out-of-process API. There are quite a few other considerations, specifically pertaining to non-functional requirements (such as performance, resilience, error handling, and more). We will look at these in more detail in *Chapter 11, Decomposing into Finer-Grained Components*.

Now that we have a handle on how we can work with APIs that interact with the frontend, let's look at how we can handle event publication and consumption *remotely*.

Changes for event interactions

Currently, our application publishes and consumes domain events over an in-process bus that the **Axon** framework makes available.

We publish events when processing commands:

```
class LCApplication {

    // Boilerplate code omitted for brevity
    @CommandHandler
    public LCApplication(StartNewLCApplicationCommand command) {
        //...
        AggregateLifecycle.apply( ❶
                new LCApplicationStartedEvent(command.getId(),
                command.getApplicantId(), command.getClientReference(),
LCState.DRAFT));
    }
}
```

1. Publishing an event when processing a command successfully and consume events to expose query APIs:

```
1  class LCApplicationSummaryEventHandler {
2
3      // Boilerplate code omitted for brevity
4
5      @EventHandler    ❶
6      public void on(LCApplicationStartedEvent event) {
7          //...
8      }
9  }
```

1.   We subscribe to an event using the Axon-provided @EventHandler annotation.

In order to process events remotely, we need to introduce an explicit infrastructure component in the form of an event bus. Common options include message brokers such as **ActiveMQ** and **RabbitMQ**, or a distributed event streaming platform, such as **Apache Kafka**. Application components can continue to publish and consume events as before—only, now, they will happen using an out-of-process style of invocation. Logically, this now causes our application to look similar to the following:

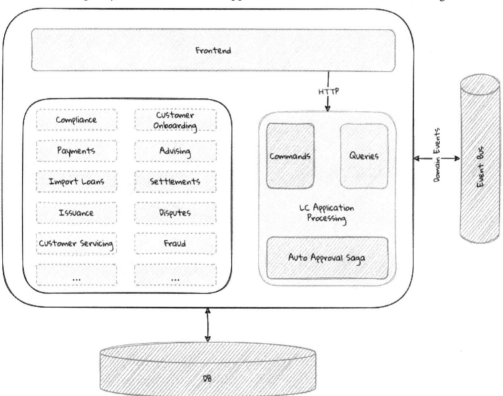

Figure 10.3 – Introducing an out-of-process event bus

When working with events within the confines of a single process, assuming synchronous processing (event publishing and consumption on the same thread), we do not encounter a majority of problems that only become apparent when the publisher and the consumer are distributed across multiple processes. Let's examine some of these in more detail next.

Atomicity guarantees

Previously, when the publisher processed a command by publishing an event and the consumer(s) handled it, transaction processing occurred as a single atomic unit, as follows:

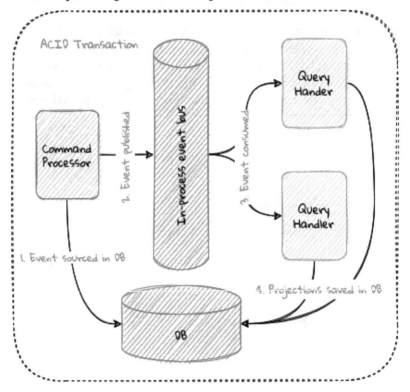

Figure 10.4 – ACID transaction processing within the monolith

Notice how all the highlighted operations in the preceding diagram happen as part of a single database transaction. This allows the system to be strongly consistent from end to end. When the event bus is distributed to work within its own process, atomicity cannot be guaranteed as it was previously. Each of the preceding numbered operations works as an independent transaction. This means that they can fail independently, which can lead to data inconsistencies.

To solve this problem, let's look at each step of the process in more detail, starting with command processing:

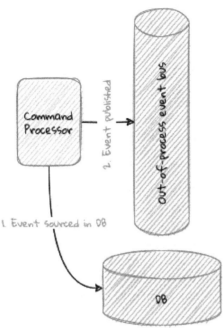

Figure 10.5 – Command processing transaction semantics

Let's consider a situation where we save to the database but fail to publish the event. Consumers will remain oblivious of the event that is occurring and become inconsistent. On the flip side, if we publish the event but fail to save it in the database, the command processing side itself becomes inconsistent—not to mention that the query side now thinks that a domain event occurred, when, in fact, it did not. Again, this leads to inconsistency. This **dual-write** problem is fairly common in distributed event-driven applications. If command processing has to work in a foolproof manner, saving to the database and publishing to the event bus have to happen atomically—both operations should succeed or fail in unison. Here are a few solutions that we have used to solve this issue (in increasing order of complexity):

- **Do nothing**: Arguably, this approach is not really a solution; however, it might be the only placeholder until a more robust solution is in place. While it might be puzzling to see this being listed as an option, we have seen several occasions where this is indeed how event-driven systems have been implemented. We leave this here as a word of caution so that teams become cognizant of the pitfalls.

- **Transaction synchronization**: In this approach, multiple resource managers are synchronized in a way that a failure in any one system will trigger a cleanup in the others where the transaction has already been committed. It is pertinent to note that this might not be foolproof, as it might lead to cascading failures.

Information

The Spring Framework provides support for this style of behavior through the `TransactionSynchronization` interface and the now deprecated `ChainedTransactionManager` interface. Please refer to the framework documentation for further details. Needless to say, this interface should not be used without careful consideration of the business requirements.

- **Distributed transactions**: Another approach is to make use of distributed transactions. A distributed transaction is a set of operations on data that is performed across two or more resource managers (usually, these are databases) using techniques such as **two-phase commit** (`https://martinfowler.com/articles/patterns-of-distributed-systems/two-phase-commit.html`). Typically, this functionality is implemented using pessimistic locking on the underlying resource managers (databases) and could present scaling challenges in highly concurrent environments.

- **Transactional outbox**: None of the preceding methods are completely foolproof in the sense that there still exists a window of opportunity where the database and the event bus can become inconsistent (this is true even with two-phase commits). One way to circumvent this problem is by completely eliminating the dual-write problem.

  In this solution, the command processor writes to its database and the intended event to an *outbox* table in a local transaction. A separate poller component polls the outbox table and writes to the event bus. Polling can be computationally intensive and could lead back to the dual write problem again because the poller has to keep track of the last written event. This could be avoided by making event processing idempotent on the consumer so that processing duplicate events do not cause issues, especially in extremely high concurrency and volume scenarios. Another way to mitigate this issue is to use a **Change Data Capture** (**CDC**) tool (such as **Debezium**, `https://debezium.io/`) and Oracle LogMiner (`https://en.wikipedia.org/wiki/OracleLogMiner`). Most modern databases ship with tools to make this easier, and they may be worth exploring. One way to implement this is to use the **transactional outbox pattern**, as shown in the following diagram:

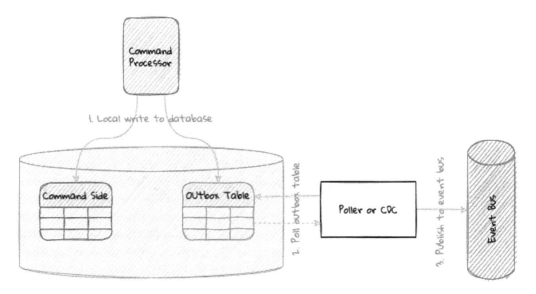

Figure 10.6 – The transactional outbox pattern

The transactional outbox pattern is a robust approach for dealing with the dual-write problem. However, it also introduces a non-trivial amount of operational complexity. In one of our previous implementations, we made use of transactional synchronization to ensure that we never missed writes to the database. Additionally, we ensured that the event bus was highly available through redundancy on both the compute and storage tiers and, most importantly, by avoiding *any* business logic on the event bus.

Delivery guarantees

Previously, because all of our components worked within a single process, the delivery of events to the consumers was guaranteed at least as long as the process stayed alive. Even if event processing failed on the consumer side, it was fairly straightforward to detect the failure because exception handling was fairly straightforward.

Furthermore, rollbacks were straightforward because the production and consumption of events happened as part of a single database transaction. With the LC processing application now becoming a remote component, event delivery becomes a lot more challenging. When it comes to message delivery semantics, there are three basic categories:

- **At-most-once delivery**: This means that each message might be delivered once or not at all. Arguably, this style of delivery is the easiest to implement because the producer creates messages in a fire-and-forget fashion. This might be okay in environments where the loss of some messages is tolerated. For example, data from click-stream analytics or logging can fall into this category.

- **At-least-once delivery**: This means that each message will be delivered more than once with no messages being lost. Delivery of undelivered messages is retried—potentially infinitely. This style of delivery might be required when it is not feasible to lose messages, but where it would be tolerable to process the same message more than once. For example, analytical environments can tolerate duplicate message delivery or have duplicate detection logic to discard already processed messages.

- **Exactly-once delivery**: This means that each message is delivered exactly once without either being lost or duplicated. This style of message delivery is extremely hard to implement, and a lot of solutions could approach exactly-once semantics with some implementation help from the consumers where duplicate messages are detected and discarded with the producer sticking to at-least-once delivery semantics.

For domain event processing, most teams will, of course, prefer to have exactly-once processing semantics, given that they would not want to lose any of these events. However, given the practical difficulties guaranteeing *exactly-once* semantics, it is not unusual to approach exactly-once processing by having the consumer process events in an idempotent manner or designing events to make it easier to detect errors.

For example, consider a `MonetaryAmountWithdrawn` event, which includes `accountId` and `withdrawalAmount`. This event could carry an additional `currentBalance` attribute so that the consumer will know if they are out of sync with the producer when processing the withdrawal. Another way to do this is for the consumer to keep track of the last *n* events processed. When processing an event, the consumer can check whether this event has already been processed. If so, they can detect it as a duplicate and simply discard it. Again, all the preceding methods add a level of complexity to the overall system. Despite all these safeguards, consumers could still find themselves out of sync with the system of record (the command side that produces the event). If so, as a last resort, it might be necessary to use partial or full event replays (`https://ddd-book.karthiks.in/10-distributing-into-multiple-components.html#_historic_event_replays`), which was discussed in *Chapter 7, Implementing Queries*.

Ordering guarantees

In an event-driven system such as the one we are building, it is desirable for consumers to receive events in a deterministic order. Not knowing the order or receiving it in the wrong order could result in inaccurate outcomes. Let's consider the example of `LCApplicationAddressChangedEvent` occurring twice in *quick succession*. If these changes are processed in the wrong order, we could end up displaying the wrong address as their current one. This does not necessarily mean that events need to be ordered for all use cases. Let's consider another example where we receive `LCApplicationSubmittedEvent` more than once erroneously when it is not possible to submit a given LC application more than once. All such notifications after the first can be ignored.

As a consumer, it is important to know whether events will be ordered or not so that we can make design considerations for out-of-order events. One default might be to accommodate for out-of-order events as a default. In our experience, this does tend to make the resulting design more complicated, especially in cases where the order does matter. Here, we will discuss three-event ordering strategies and their implications for both the producer and the consumer:

Strategy	Producer	Event bus	Consumer
No ordering	Arguably, this is the easiest to implement because there is no expectation from the producer to support ordering.	Without additional metadata, the event bus might only be able to guarantee ordering in the sequence of a receipt (**FIFO** order).	If the consumer depends on ordering, it might have to implement ordering through some form of special processing.
Per aggregate ordering	The producer needs to make sure that each event includes an identifier to enable grouping by aggregate.	The event bus needs to support the notion of grouping (in this case, by the aggregate identifier). For events belonging to the same aggregate instance, messages are emitted in FIFO order.	To guarantee ordering, events originating from the same aggregate instance need to be processed by the same consumer instance.
Global ordering	The producer needs to make sure that each event includes a sequence number.	Either the event bus or the consumer needs to implement ordering logic.	

In most applications, per aggregate ordering might be a good place to start and cater to most business scenarios.

Durability and persistence guarantees

When an event is published to the event bus, the happy path scenario is that the intended consumer(s) can process it successfully. However, there are scenarios that can cause message processing to be impacted adversely. Let's examine each of these scenarios:

- **Slow consumer**: The consumer is unable to process events as fast as the producers are publishing them.

- **Offline consumer**: The consumer is unavailable (down) at the time of the events being published.

- **Failing consumer**: The consumer is experiencing errors when trying to process events.

In each of these cases, we could develop a backlog of unprocessed events. Because these are domain events, we need to prevent the loss of these events until the consumer has been able to process them successfully. There are two communication characteristics that need to be true for this to work successfully:

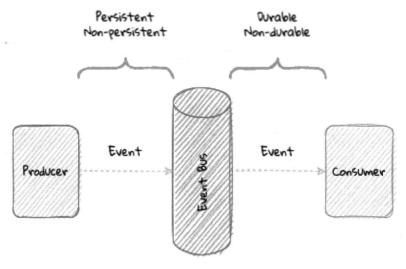

Figure 10.7 – Persistence versus durability

- **Persistence**: This is the communication style between the **Producer** instance and the **Event Bus** instance.

- **Durability**: This is the communication style between the **Event Bus** instance and the **Consumer** instance.

Firstly, messages need to be persistent (that is, stored on disk), and secondly, the message subscription (the relationship between the consumer and the event bus) needs to be durable (persist across **Event Bus** restarts). It is important to note that events have to be made persistent by the producer for them to be consumed durably by the consumer.

Processing guarantees

When an event is processed by the query side component, as shown here, the following steps occur:

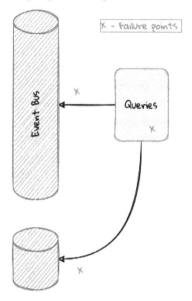

Figure 10.8 – Event processing failure scenarios

1. The event is consumed (either through a push or a pull) from the **Event Bus** instance.
2. Transformation logic is applied to the payload of the event.
3. The transformed payload is saved in the query side store.

Each of these steps can encounter failures. Irrespective of the cause of failure, the event should be durable (as discussed earlier) so that it can be processed later when the issue is fixed. These errors can be broadly segregated into four categories:

Error cause	Example	Remediation
Transient	A network blip, resulting in temporary connectivity issues to either the event bus or the query store.	A finite number of retries, potentially with a backoff strategy, before giving up with a fatal error.
Configuration	Event bus or database URL misconfiguration.	Manual intervention with updated configuration and/or restart.
Code logic	Implementation bugs in the transformation logic.	Manual intervention with updated logic and redeployment.
Data	Unexpected or erroneous data in the event payload.	Manual intervention that requires segregating spurious data (for example, by automatically moving problematic events to a **dead-letter queue**) and/or fixing code logic.

Now we have looked at the changes that we need to make because of the introduction of an out-of-process event bus. Having done this allows us to actually extract the **LC Application Processing** component into its own independently deployable unit, which will look similar to the following diagram:

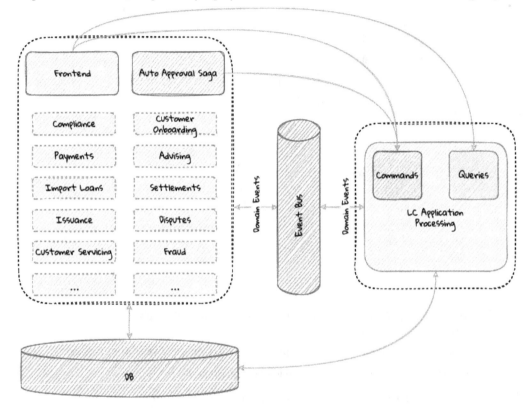

Figure 10.9 – LC Application Processing deployed independently

However, we are continuing to use a common datastore for the **LC Application Processing** component. Let's look at what is involved in segregating this into its own store.

Changes in database interactions

While we have extracted our application component into its own unit, we continue to be coupled at the database tier. If we are to achieve true independence from the monolith, we need to break this database dependency. Let's look at the changes involved in making this happen.

Data migration

As a first step to start using a database of our own, we will need to start migrating data from the command side event store and the query store(s), as shown here:

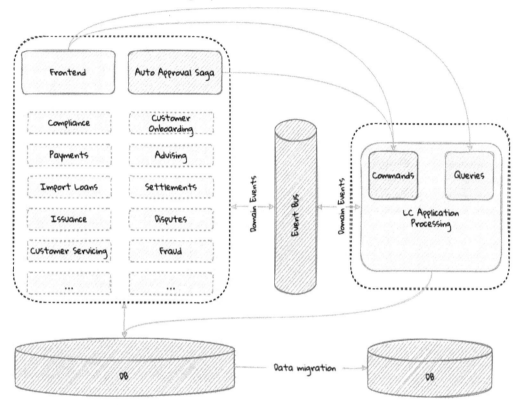

Figure 10.10 – Data migration

In our case, we have the command side event store and the query store(s) that will need to be migrated out. To minimize effort from the outset, it might be prudent to do a simple homogenous migration by keeping the source and target database technologies identical. In advance of the cut-over, among other things, it will be essential to do the following:

- **Profile** to make sure that latency numbers are within tolerable limits
- **Test** to make sure that the data has migrated correctly
- **Minimize downtime** by understanding and agreeing on **SLAs**, such as the **Recovery Time Objective (RTO)** and **Recovery Point Objective (RPO)**

Cut-over

If we have made it so far, we are ready to complete the migration of the LC Application Processing from the rest of the monolith. The logical architecture of our solution now looks similar to the following diagram:

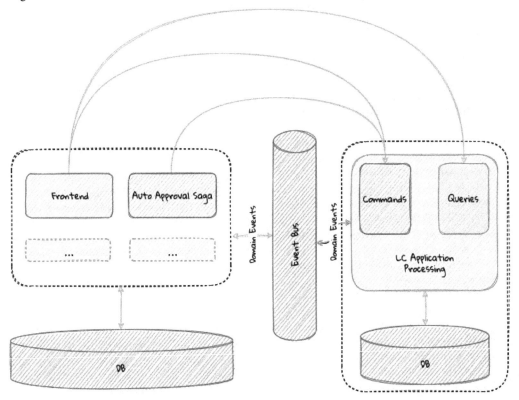

Figure 10.11 – Independent data persistence

With this step, we have successfully completed the migration of our first component. There is still quite a lot of work to do. Arguably, our component was already well-structured and loosely coupled with the rest of the application. Despite that, moving from an in-process model to an out-of-process model between bounded contexts is quite an involved process—as should be evident from the work we have done in this chapter.

Summary

In this chapter, we learned how we can extract a bounded context from an existing monolith, although you could argue that this was from a reasonably well-structured one. We looked at the challenges involved in decomposing the monolith from various interaction points such as the frontend, event exchanges, and the database. You should have an understanding of what it takes to go from an in-process event-driven application to an out-of-process one.

In the next chapter, we will look at how to extract pieces out of a monolith that might not be as well-structured, possibly very close to the dreaded big ball of mud.

References

For more information, please refer to the following resources:

- `https://roy.gbiv.com/untangled/2008/rest-apis-must-be-hypertext-driven`
- `https://martinfowler.com/articles/patterns-of-distributed-systems/two-phase-commit.html`
- `https://debezium.io/`

11
Decomposing into Finer-Grained Components

In the previous chapter, we decomposed the *LC Application Processing* functionality out of the monolith. In this chapter, we will further decompose these components into even more fine-grained components. In addition, we will examine if and when such a decomposition is justified.

The following topics will be covered in this chapter:

- Continuing our design journey
- Even more fine-grained decomposition
- Decomposing the frontend
- Where to draw the line

At the end of this chapter, you will be able to appreciate both technical and non-technical factors that play toward where we should draw the line on decomposing these components.

Continuing our design journey

Currently, our application resembles the diagram depicted here:

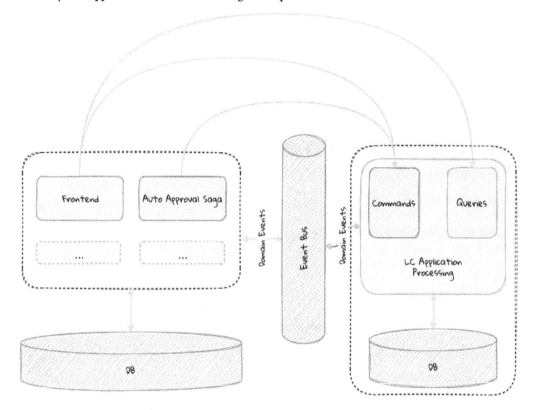

Figure 11.1 – Independent data persistence

The *LC Application Processing* functionality lives as its own independent component from the rest of the application. It communicates with the monolith through the exchange of domain events using the event bus. It makes use of its own persistence store and exposes HTTP-based APIs that the frontend consumes. Let's examine whether it is possible to further decompose the application into finer-grained components. The AutoApprovalSaga component currently lives within the confines of the monolith, but this is mostly an artifact of our previous design as opposed to an intentional design choice. Let's look at how we can extract this into its own component next.

Saga as a standalone component

Currently, the `AutoApprovalSaga` component (discussed in detail in *Chapter 8*, Implementing *Long-Running Workflows*) works by listening to domain events, as shown here:

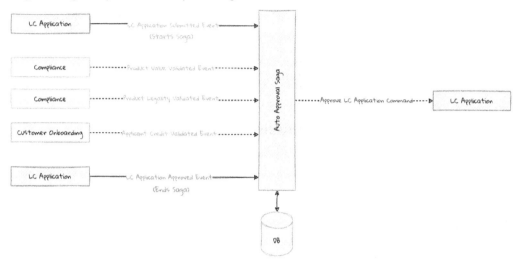

Figure 11.2 – The AutoApprovalSaga functionality dissected

Given that these events are published by different bounded contexts onto the event bus, there is no need for `AutoApprovalSaga` to be embedded within the monolith. This means that it can be safely pulled out into its own deployable unit along with its private data store. This means that our system now looks like the diagram depicted here:

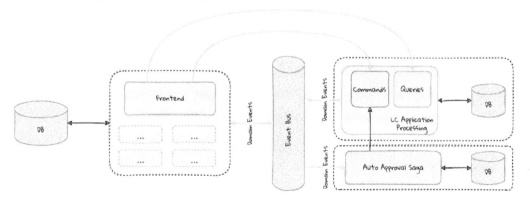

Figure 11.3 – AutoApprovalSaga extracted into an independent component

Saga components can be characterized as a collection of stateful event listeners listening to events from more than one aggregate that can issue commands to more than aggregate. We saw earlier that we form bounded contexts along aggregate boundaries. Given that sagas tend to require interaction with more than one aggregate, they may not fall within the confines of those bounded contexts. In a lot of ways, sagas are components that can be viewed as their own bounded contexts. This makes it natural to have sagas work as standalone components that exist distinctly (both from a logical and physical perspective) from other parts of a solution.

As you can see, commands and queries within the *LC Application Processing* component continue to use a common data store. Let's look at what is involved in segregating them into their own data store.

Commands and queries as standalone components

As we saw in the *CQRS pattern* section in *Chapter 2, Where and How Does DDD Fit?*, the primary benefit that we derive is the ability to evolve and scale these components independently of each other. This is important because commands and queries have completely different usage patterns and thus require the use of distinct domain models. This makes it fairly natural to further split our bounded contexts along these boundaries. Thus far, the segregation is logical. A physical separation will enable us to truly scale these components independently, as shown here:

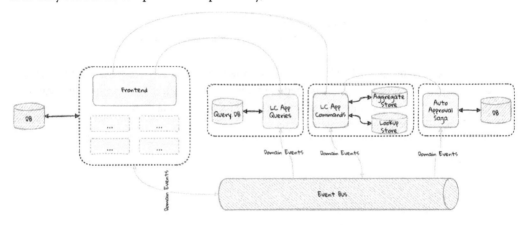

Figure 11.4 – Commands and queries as independent components

It is pertinent to note that the command processing component is now shown to have access to two distinct data stores:

- The **aggregate store**, which stores either an event-sourced or state-stored representation of an aggregate state.

- The **lookup store**, which can be used to store lookup data when performing business validations when processing commands. This is applicable when we need to access data that is/cannot be stored as part of the aggregate state.

The reason we bring this up is that we may have to continue making lookups for data that still remains in the monolith. To achieve full independence, this lookup data must also be migrated using techniques such as a historic event replay (as discussed in *Chapter 7, Implementing Queries*) or other conventional data migration techniques (as discussed in *Chapter 10, Beginning the Decomposition Journey*).

Distributing individual query components

At this point, we have achieved segregation along command and query boundaries. But we do not need to stop here. Each of the queries we service need not necessarily remain a single component. Let's consider an example where we need to implement a fuzzy LC search feature for the UI and a view of LC facts for analytical use cases. It is conceivable that these requirements may be implemented by a different set of teams, thereby necessitating the need for distinct components. Even if these are not distinct teams, the disparity in usage patterns may warrant the use of different persistence stores and APIs, again requiring us to look at implementing at least a subset of these as distinct components, as shown here:

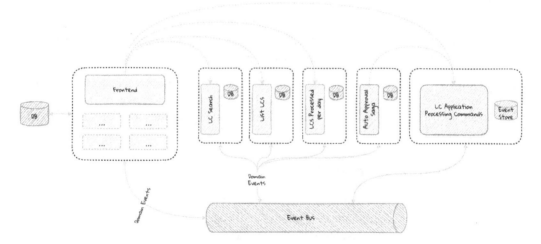

Figure 11.5 – Queries split into individual components

Owning domains should strive to create query APIs that exhibit the characteristics of a good domain data product. Some of these characteristics include being discoverable, trustworthy, valuable in their own right, and self-describing. For more information, please refer to this article on moving from a monolithic data lake to a distributed data mesh. Specifically, the section on domain data as a product is relevant in this context: `https://martinfowler.com/articles/data-monolith-to-mesh.html#DomainDataAsAProduct`.

Even more fine-grained decomposition

At this stage, is there any further decomposition that is required and feasible? These days, whether rightfully or otherwise, serverless architecture (specifically, *functions as a service*) is arguably becoming all the rage. As we pointed out in *Chapter 2, Where and How Does DDD Fit?*, this means that we may be able to decompose our command side in a manner that each command becomes its own independently deployable unit (hence a bounded context). In other words, `LCApplicationSubmitCommand` and the `LCApplicationCancelCommand` can be deployed independently.

But just because this is technically possible, should we do it? While it is easy to dismiss this as a passing fad, there may be good reasons to split applications along command boundaries:

- **Risk profile**: Certain pieces of functionality present a higher risk when changes are made. For example, submitting an LC application may be deemed a lot more critical than the ability to cancel it. However, that is not to say that *canceling* is unimportant. Being decoupled from *submit* allows *cancel* changes to be made with a lot less scrutiny. This may make it easier to innovate quickly with more experimental features, with minimal fear of causing large disruptions.

- **Scalability needs**: Scaling needs can differ wildly for various commands in a system. For example, *submit* may need to scale a lot more than *cancel*. However, being coupled will force us to treat them as equals, which can be inefficient.

- **Cost attribution**: Having fine-grained components allows us to more accurately measure the amount of effort and the resulting ROI dedicated to each individual command. This can make it easier to focus our efforts on the most critical functionality (the "core" of the core) and minimize waste.

Effects on the domain model

These finer grainer components are leading us to a point where it may appear that the deployment model is starting to have a big influence on the design. The fact that it is now feasible to deploy individual "tasks" independently requires us to reexamine how we arrive at bounded contexts. For example, we started by working on the *LC Application Processing* bounded context, and our aggregate design was based on all functionality included in the scope of application processing. Now, our aggregate design can be a lot more fine-grained. This means that we can have an aggregate specifically for *start* functionality and another for *cancel*, as shown here:

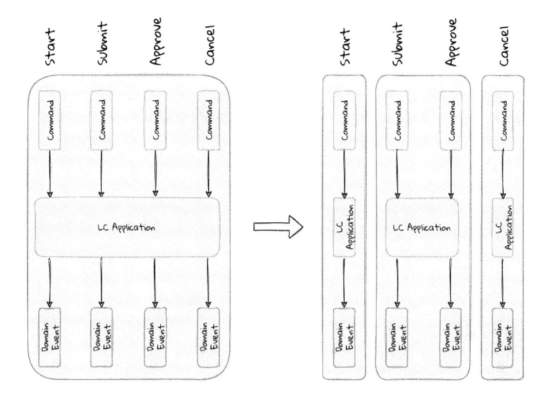

Figure 11.6 – A fine-grained bounded contexts example

The most fine-grained decomposition may lead us to a bounded context per command, but that does not necessarily mean that we have to decompose the system this way. In the preceding example, we have chosen to create a single bounded context for the *submit* and *approve* commands. However, *start* and *cancel* have their own bounded contexts. The actual decision that you make in your own ecosystems will depend on maintaining a balance among reuse, coupling, transactional consistency, and other considerations that we discussed earlier. It is important to note that the aggregate labeled as LCApplication, although named identically, is distinct from a domain model perspective in its respective bounded context. The only attribute they will need to share is a **common identifier**. If we choose to decompose the system into a bounded context per command, our overall solution will look like the diagram shown here:

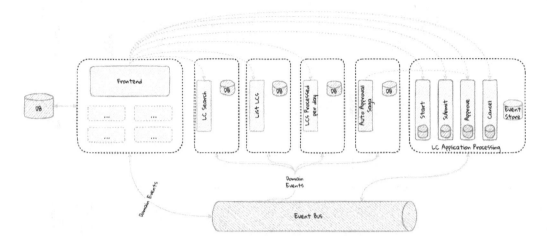

Figure 11.7 – Decomposition per command

It is pertinent to note that the *command* functions continue to share a single event store, although they may make use of their own individual lookup stores. We understand that this decomposition likely feels unnecessary and forced. However, this does allow us to focus our energies on the *core of the core*. For example, LC application processing may be our business differentiator. However, an even more careful examination may reveal that it is our ability to *decision* LCs near real time that is our real business differentiator. This means that it may be prudent to isolate that functionality from the rest of the system. In fact, doing so may enable us to optimize our business process without adding risk to the overall solution. While it is not strictly necessary to decompose the system in this way to arrive at such insights, a fine-grained decomposition may enable us to refine the idea of what is most important to our business. Having to share a persistent store can be a wrinkle to achieve complete independence. Therefore, a final decomposition may look something like the following:

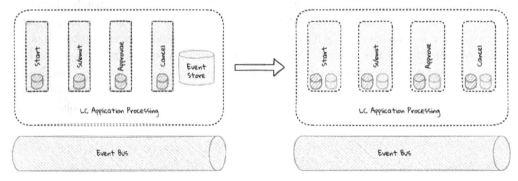

Figure 11.8 – Command components with individual event stores

Obviously, there is no free lunch! This fine-grained decomposition may require additional coordination and duplication of data among these components – to a point where it may not be attractive anymore. However, we feel that it is important to illustrate the art of the possible.

Decomposing the frontend

Thus far, we have focused on decomposing and distributing the backend components while keeping the frontend untouched as part of the existing monolithic system. It is worth considering breaking down the frontend to align it more closely along functional boundaries. Patterns such as micro-frontends (`https://micro-frontends.org/`, `https://martinfowler.com/articles/micro-frontends.html`) extend the concepts of microservices to the frontend. Micro-frontends promote team structures to support end-to-end ownership of a set of features. It is conceivable that a cross-functional, polyglot team owns both the experience (frontend) and the business logic (backend) functions, eliminating communication overheads drastically (along the lines of the vertical slice architecture conversation, as discussed in *Chapter 2, Where and How Does DDD Fit?*). Even if such a team organization where the frontend and backend are one team is not feasible in your current ecosystem, this approach still has many merits, such as the following:

- **Increased end-to-end collaboration**: Creating solutions that work end to end is what ultimately provides value. Having a set of backend services isolated from their respective customer experiences will only cause us to accumulate unused inventory. To reduce the possibility of failure, the closer the collaboration between the backend capability and frontend experience teams, the greater our chance of reducing waste due to misaligned requirements. Including the customer experiences as part of the vertical slice allows us to apply the ubiquitous language through the entire stack.

- **Uniform omnichannel experiences**: These days, it is very common to surface the same functionality across more than one experience channel. Having an inconsistent experience across channels can lead to customer dissatisfaction and/or adverse business consequences. Aligning teams closely along functional boundaries (within the same *swim lane*) can promote high levels of collaboration and consistency when exposing business functionality. Consider the example shown here. Within a vertical slice, the allegiance is to the functionality being developed, although there may be a need to use disparate technologies to build each channel (iOS, Android, web, and so on). Within a vertical slice, each box depicted in the diagram may operate as a team of its own, while maintaining strong cohesion with the functional team within the same swim lane, as shown here:

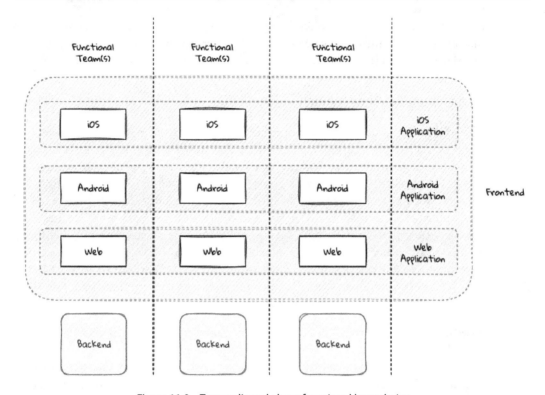

Figure 11.9 – Teams aligned along functional boundaries

While there are many advantages in employing this approach, as with everything else, it does come with a few gotchas that you may need to be mindful of:

- **End-to-end testing complexity**: While this is true for a lot of distributed architectures, this problem is exacerbated in the case of user experiences because of it being a visual medium. Especially if real components come together close to the end of the cycle, it may become harder to visualize the end-to-end flow until almost all the visual elements are in place. This may also be in conflict with how end users interact with a system as a whole. This may make end-to-end testing complex because it needs components from multiple teams to come together, possibly close to the end of the cycle.

- **Deployment complexity**: In the preceding example, we have split the application along functional boundaries. However, they have to come together as a single artifact at the time of deployment (this is especially true in the case of mobile applications). This can add quite a bit of deployment complexity when the complete application is assembled. It is important to be cognizant of the relationship patterns between teams (as covered in *Chapter 9, Integrating with External Systems*) to work through kinks.

- **Dependency management**: Given that teams may need to ultimately deploy the application as a unit, managing dependencies between individual modules may become cumbersome. This may manifest itself in the form of conflicting dependency versions, leading to unpredictable and inefficient runtime behavior and performance. For example, two teams may use different versions of the same frontend library, which may add to the overall payload that gets downloaded to the browser. In addition to being wasteful, this may also result in unpredictable, hard-to-diagnose errors, and eventually, poor customer experience.

- **Inconsistent user experiences**: Although we may have split the application in a seemingly logical manner, if we don't do it in a manner that is transparent to the end users, it may result in a confusing and likely frustrating experience. To mitigate this, there may be a need to build common assets, widgets, and so on. which may further add to the overall complexity and coordination required when shipping out the end product.

If we proceed to continue decomposing our application as suggested previously, our application will end up looking like the diagram shown here:

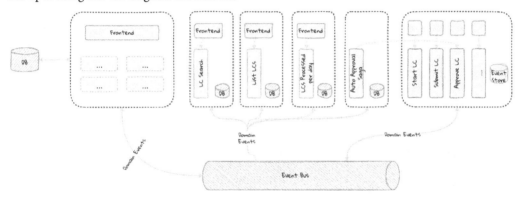

Figure 11.10 – Commands and query frontends decomposed into individual functions

As we saw, there are multiple ways to approach decomposing an application into finer-grained components. Just because it is possible to do it, it doesn't mean that we should. Let's look at when decomposition starts to become too expensive to sustain productivity.

Where to draw the line

In general, the smaller the size of our bounded contexts, the easier it becomes to manage complexity. Does that mean we should decompose our systems into as fine-grained a granularity as possible? On the other hand, having extremely fine-grained components can increase coupling among them to the extent where it becomes very hard to manage operational complexity. Hence, decomposing a system into well-factored, collaborating components can be a bit tricky, seeming to work more like an art rather than an exact science. There is no right or wrong answer here. In general, if things feel and become painful, you most likely got it more wrong than right. Here are some non-technical heuristics that might help guide this process:

- **Existing organization boundaries**: Look to align along with current organizational structures. Identify which applications your business unit/department/team already owns and assign responsibilities in a manner that causes minimal disruption.

- **End-user roles and responsibilities**: What work do your end users carry out? What enables them to do their work with the least friction possible? If too many people need to get involved to get a piece of work done, that may be a sign that the current decomposition may be suboptimal. On the other hand, if it is hard to assign a task to a specific user, it may again be a sign of incorrect decomposition.

- **Change in vernacular**: Look for subtle changes in the usage of common terms (the *ubiquitous language*). Does someone call something that is/feels the same in the physical world by different names? For example, a credit card can be called "plastic," "payment instrument," and "account" by different people or the same people in a different context. The point at which the vernacular changes may be the time to split functionality.

- **Existing (modular/monolithic/distributed) applications**: How are your current applications segregated logically? How are they segregated physically? This might provide some inspiration.

Team organization

All of the preceding techniques draw inspiration from existing constructs. However, what if one or more of the preceding are wrong/cumbersome/suboptimal? In such a case, our work as developers/ architects is a bit more involved.

It is also pertinent to note that it is not uncommon to get domain boundaries wrong. Coming up with an initial breakdown that seems to make more sense and applying a series of *what if* questions to assess suitability can help. If the reasoning is able to stand up to scrutiny by domain experts, architects, and other stakeholders, you might be in a good place. If you do choose to go down this route, it may be prudent to adjust existing organizational structures to match your proposed architecture. This will help reduce friction (in other words, apply what is called the *inverse Conway maneuver* (`https://www.thoughtworks.com/en-us/radar/techniques/inverse-conway-maneuver`).

This style of team organization can be quite complex. The people at Spotify popularized the idea of a multidisciplinary, mostly autonomous team structure aligning closely along functional boundaries (called *squads*), as shown here:

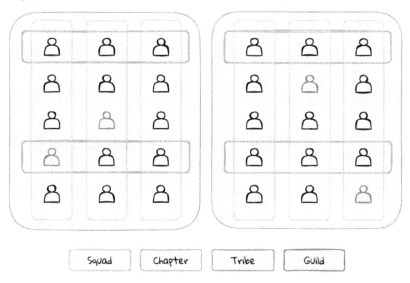

Figure 11.11 – The Spotify model of team organization

The team structure has other components such as chapters, tribes, and guilds, which enable a better flow of change, clarify team responsibilities, promote better intra- and inter-team collaboration, and so on. You can find out more about it in this post: https://blog.crisp.se/wp-content/uploads/2012/11/SpotifyScaling.pdf. However, there is no one-size-fits-all approach, and you will need to account for your own organizational structures and realities before looking to adopt this style. To find out more about the *limitations of the Spotify model* (https://www.youtube.com/watch?v=4GK1NDTWbkY) and how you can arrive at a team organization that better suits your own requirements, you may want to take a look at the work done by Matthew Skelton and Manuel Pais in their popular book *Team Topologies* (https://teamtopologies.com/book). On a related note, it may also be helpful to look at the chapter on team design from the book *Agile IT Organization Design* (https://www.amazon.com/Agile-Organization-Design-Transformation-Continuous/dp/0133903354) by Sriram Narayan, where he talks about outcome-oriented versus activity-oriented teams.

Despite all our due diligence and noble intentions, it is still possible to get these boundaries wrong, or a change in business priorities or competitor offerings may render decisions that appeared perfectly valid at the time to become incorrect. Instead of looking to arrive at the perfect decomposition, it might be prudent to embrace change and invest in building designs that are flexible while being prepared to evolve and refactor the architecture iteratively. This book on building evolutionary architectures has some great advice on how to do precisely that: https://evolutionaryarchitecture.com/.

In order to attain a reasonable level of success, there will be a need to maintain a fine balance between how domains are modeled, what the team organizations are, and how applications are architected. When all of these are in agreement, it is likely that you get pretty close to achieving high levels of success, as depicted in the following diagram:

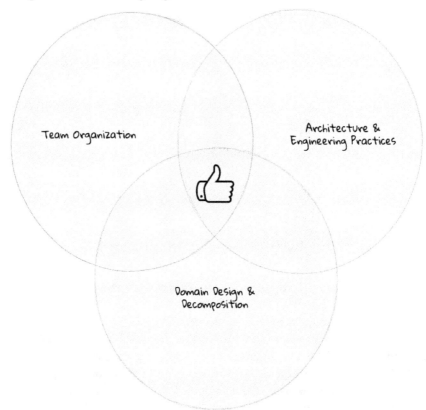

Figure 11.12 – Forces influencing component decomposition

As a general guideline, it helps to start with a coarse-grained decomposition at the outset when requirements and/or our understanding are likely still unclear, leaving finer-grained decomposition to a time when our understanding improves.

Summary

In this chapter, we learned how an already fine-grained application can be further decomposed to the level of individual functions, each of which may be deployed as its own independent unit. We looked at how we stand to benefit from keeping end-to-end functionality (a thin vertical slice) as a cohesive unit, which includes components from the frontend experience all the way to the backend.

Further, we looked at how Conway's law can play an important role in the evolution of our architecture. We also looked at how we may be able to course correct cumbersome organizational structures by applying the inverse Conway maneuver. Finally, we briefly touched on popular methods of team organization that you can take inspiration from when designing your own organizational structures.

In the next chapter, we will look at a variety of non-functional characteristics that play a significant role in how we can decompose and distribute applications.

Further reading

Title	Author	Location
How to Move Beyond a Monolithic Data Lake to a Distributed Data Mesh	Zhamak Dehghani	`https://martinfowler.com/articles/data-monolith-to-mesh.html#DomainDataAsAProduct`
Micro Frontends	Michael Geers	`https://micro-frontends.org/`
Micro Frontends	Cam Jackson	`https://martinfowler.com/articles/micro-frontends.html`
Inverse Conway Maneuver	Thoughtworks Tech Radar	`https://www.thoughtworks.com/en-us/radar/techniques/inverse-conway-maneuver`
Scaling Agile @ Spotify with Tribes, Squads, Chapters & Guilds	Henrik Kniberg and Anders Ivarsson	`https://blog.crisp.se/wp-content/uploads/2012/11/SpotifyScaling.pdf`
Spotify Engineering Culture – Part 1	Henrik Kniberg	`https://www.youtube.com/watch?v=4GK1NDTWbkY`
Spotify Engineering Culture – Part 2	Henrik Kniberg	`https://www.youtube.com/watch?v=vOt4BbWLWQw`
Team Topologies	Matthew Skelton and Manuel Pais	`https://teamtopologies.com/book`
Agile IT Organization Design	Sriram Narayan	`https://www.amazon.com/Agile-Organization-Design-Transformation-Continuous/dp/0133903354`
Building Evolutionary Architectures	Rebecca Parsons, Neal Ford, and Pat Kua	`https://evolutionaryarchitecture.com/`

12
Beyond Functional Requirements

Sometimes I feel like I am being forgotten.

— *Anonymous*

While the functional requirements of the core of the system may be met adequately, it is just as important to place focus on the operational characteristics of the system. In this chapter, we will look at common pitfalls and how to get past them.

In this chapter, we will cover the following topics:

- Observability
- Consistency
- Performance and scale
- Trunk-based development
- Continuous testing
- Deployment automation
- Refactoring
- Invocation style
- Logging
- Versioning

By the end of this chapter, we will have learned about various aspects of the software life cycle to create a robust solution from a cross-functional perspective. We will also discuss additional features that we will need to add to make our solution performant, scalable, resilient to failure, and gain the ability to make changes reliably, repeatably, and rapidly. Furthermore, we will also examine the implications of making these changes and the potential impacts this may have on our bounded contexts and their boundaries.

Let's begin!

Observability

In previous chapters, we saw how it is possible to break down an existing application along bounded context boundaries. We also saw how it is possible to split bounded contexts to be extremely fine-grained, often as physically disparate components. Failure in any of these components can cause disruptions in others that are dependent on them. Obviously, early detection and more importantly attribution to specific components through a combination of proactive and reactive monitoring can ideally prevent or, at the very least, minimize business disruption.

When it comes to monitoring, most teams seem to think of **technology runtime metrics** that we associate with components (such as CPU utilization, memory consumed, queue depths, exception count, and so on).

Lending Objectivity to Metrics

To make it more formal, we use the terms **Service-Level Objectives (SLOs)** and **Service-Level Indicators (SLIs)** specified within a **Service-Level Agreement (SLA)** to mean the following:

- **SLO**: An agreement between the provider and customer about a specific measurable metric. For example, 99.99% uptime, 100 ms response time for 1,000 concurrent users for requests in the 99th percentile, and so on.

- **SLA**: A collection of SLOs.

- **SLI**: The actual numbers against an SLO. For example, your system might have an uptime SLI of 99.95%.

Technology metrics

When it comes to monitoring, most teams seem to think of technology runtime metrics that we associate with components (such as CPU utilization, memory consumed, queue depths, exception count, and so on).

However, it is just as much if not more important to be able to associate a set of business-relevant metrics (such as the number of LC applications submitted in the last hour, the number of LC applications rejected, and so on) and DevOps metrics (such as lead time, mean time to restore, and so on).

Business metrics

An inability to associate and monitor business SLIs with a component may be an indicator of the component being too fine-grained. On the flip side, if there are too many business SLIs associated with a single component that is of interest to a multitude of business stakeholder groups, it may be an indicator that a more fine-grained decomposition may be justified. At the end of the day, the monitoring apparatus we have in place should be able to tell us if we are violating/meeting/exceeding SLOs.

DevOps metrics

The **DevOps Research and Assessment (DORA)** research foundation has published an online quickcheck (`https://www.devops-research.com/quickcheck.html`) tool and report (`https://www.devops-research.com/research.html`) to quickly provide information on how organizations compare with industry peers and how to make progress toward elite status. While discussing the full nuance of what it takes to establish a long-term culture of continuous improvement is out of scope for this book, we reference the four key metrics highlighted in the research paper as indicators of software delivery performance:

- **Lead time**: How long does it take to go from code committed to code successfully running in production?

- **Deployment frequency**: How often does your organization deploy code to production or release it to end users?

- **Time to restore**: How long does it generally take to restore service when a service incident or a defect that impacts users occurs?

- **Change failure percentage**: What percentage of changes to production or releases to users results in degraded service?

When it comes to observability, there is the risk of focusing on specific metrics in isolation and missing the forest for the trees. To avoid metrics being misused and, more importantly, running the risk of drawing incorrect conclusions, we recommend the following:

- **Take a holistic view**: Focusing more or less equally on all aspects of the delivery life cycle as opposed to focusing on just a particular area can go a long way. If you are able to include information from planning, requirements intake, development, build, test, deploy, and feedback from running production systems, then you may be able to conclude reasonably that you have a high-performing team.

- **Employ ratcheting**: Having recognized an improvement area, how do you go about setting yourself up for improvement? Setting clear, objective goals that are measurable and trackable (no pun intended) for improvement is paramount to be able to subsequently meet them. In order to ensure that there is continuous incremental improvement, ratcheting is a technique that can be employed. A ratchet is a device that resembles a wrench but is unique in that it only turns in one direction. In this context, ratcheting involves doing the following:

  I.   Set the current level as the minimum starting point.

  II.   Make a small incremental improvement in a relatively small amount of time.

  III. Re-adjust the baseline to the new level attained as part of step 2.

  IV.   If levels descend below the baseline, take stop-the-line action until baselines are restored.

  V.   Repeat from step 1.

Ratcheting allows teams to set incremental milestones as intermediate goals while moving closer to a much better place all the time.

Adopting an attitude of constant learning and incremental improvement through ratcheting as opposed to one that looks to police and penalize can go a long way toward instituting a system that can be effective.

Consistency

In the previous chapters, we have spent a lot of energy splitting our system into multiple, fine-grained independent components. For example, the LC application is submitted against the command-side component, whereas the status of the LC application is serviced by the query side. Because these are distinct components, there will be a time lag during which the two systems are not consistent with each other. So, querying the status of an LC application immediately after submitting may produce a stale response until the time that the query side processes the submit event and updates its internal state. In other words, the command side and the query side are considered to be *eventually consistent*. This is one of the trade-offs that we need to embrace when working with distributed systems.

Eric Brewer (professor emeritus of computer science at the University of California, Berkeley) formalized the trade-offs involved in building distributed systems in what is called the *CAP theorem*. The theorem postulates that distributed systems can either be highly available or consistent in the event of a network partition, not both at the same time. Given the three characteristics, consistency, availability, and partition tolerance, the theorem postulates that distributed systems can either be highly available or consistent in the event of a network partition, not both at the same time. This means that distributed applications that are expected to be highly available will have to forsake strong consistency.

This may make it appear that this is a deal-breaker, but in reality, most real-world business problems are tolerant to being eventually consistent. For example, there may be a requirement that an order cannot be canceled after it has shipped. In an eventually consistent system, there may exist a time window (albeit small) where we may allow a shipped order to be canceled. To deal with such scenarios, we may need to enhance the business process to account for these inconsistencies. For example, before issuing a refund for a canceled order, we may need to validate that the order has not physically shipped or has been returned. Even in the extreme case where we may have erroneously issued a refund for a shipped order, we can request the customer to return it before an expiry period to avoid getting charged. If the customer fails to return the order, we may charge the customer or write off the amount as lost business. Obviously, all this adds a level of complexity to the solution because we may need to account for edge conditions through a series of compensating actions. If none of this complexity is acceptable and strong consistency is non-negotiable, then shipping and order cancellation functionality will have to be part of the same bounded context.

Performance and scale

In previous chapters, we saw how it is possible and sometimes even necessary to break functionality down into fine-grained components that are physically separated from each other – requiring a network to collaborate. Let's assume that this collaboration is achieved in a loosely coupled manner – justifying the need for disparate bounded contexts from a logical perspective.

Performance is a very important SLO that is typically associated with most applications. When it comes to performance, it is essential to understand the basic terms. This is best illustrated using an example as shown here:

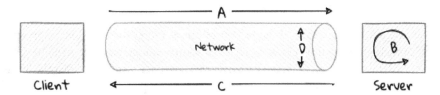

A - Amount of time taken for a request to reach the server

B - Processing time on the server

C - Amount of time taken for the response to reach the consumer

D - Maximum advertised capacity of the network

Figure 12.1 – The elements of network performance

As shown here, the following terms are relevant in the context of performance:

- **Latency**: The delay introduced by the network (A + B)
- **Response Time**: the total time taken by the system to respond to the user (A + B + C)

- **Bandwidth**: The maximum capacity of the network (D)
- **Throughput**: The amount of data processed in a given amount of time

The introduction of a network between two components introduces constraints in the form of network latency and bandwidth. Even if processing time on the server is theoretically reduced to zero, latency and bandwidth constraints cannot be avoided. This problem can only get worse as the number of network hops increases. This means that it is impossible for networked applications to provide the same level of performance as their non-networked counterparts.

The need to scale to support a larger number of requests can further complicate things. Given that Moore's law has slowed down considerably in the last decade or so, it is less feasible to continue scaling up by using more and more powerful machines. This means that beyond a point, scaling out by using multiple instances, and thereby (re-)introducing a reliance on the network, is inevitable.

This makes it evident that performance and scale requirements can have a significant impact on how we choose to distribute our components. Having a clear understanding of performance and scale SLOs is a necessary prerequisite before attempting to distribute distinct components. On the flip side, if you are in a situation where you already have distributed components that are not meeting performance and scale SLOs, one option is to aggregate them back together. If that is not feasible, it may be worth embracing alternative customer experiences along with a non-blocking, event-driven style of architecture to create a perception of better performance.

Trunk-based development

Eric Evans, the inventor of DDD, talks about how **continuous integration (CI)** helps preserve the sanctity of the domain model within a bounded context. When more than one person works in the same bounded context, it tends to fragment. Obviously, the bigger the team, the higher the likelihood of this problem occurring. Even a team as small as three or four people can encounter serious issues. We have also seen that beyond a point, there may be diminishing returns if we try to break the system into extremely fine-grained bounded contexts.

This makes it very important to institute a process of merging/integrating all code and other implementation artifacts frequently, aided by **automated tests** to flag such fragmentation. In addition, this allows the team to apply the ubiquitous language relentlessly, each time refining the domain model to represent the problem more accurately. In other words, it is critical to practice continuous integration. Many teams make use of a CI server to run tests but tend to postpone integration until very late making use of an excessive number of long-living branches (popularized by Gitflow; `https://www.atlassian.com/git/tutorials/comparing-workflows/gitflow-workflow` and merge requests–practicing an anti-pattern known as CI theatre (`https://www.gocd.org/2017/05/16/its-not-CI-its-CI-theatre.html`).

An alternative to branch-based development is *trunk-based development* where each developer works in incremental batches and merges that work into the main (also called trunk) branch at least once (and potentially several times) a day. The DORA team has published research (`https://services.`

`google.com/fh/files/misc/state-of-devops-2021.pdf#page=27`) that shows that elite performers practice trunk-based development to maximize the effectiveness of their CI practice and by extension their ability to continuously enhance their domain models and keep up with changing business needs.

In an ideal world, every commit to the trunk would constitute finished, production-ready work. But it is also fairly normal for certain pieces of work to take longer to complete. This may make it appear that there is a need to forsake trunk-based development and resort to branch-based development. However, there is no need to compromise the continuous integration flow to accommodate for such eventualities. Paul Hammant (`https://paulhammant.com/`) talks about this technique called *branch by abstraction* where the effects of unfinished pieces of work are hidden behind an abstraction layer. This abstraction layer is typically implemented by either making the new piece of functionality hidden from the end user or in more sophisticated cases, using feature flags (`https://martinfowler.com/articles/feature-toggles.html`).

Continuous testing

In an ideal world, continuous integration will enable us to adopt continuous testing, which provides us with constant and early feedback. This is essential because our bounded contexts and the resulting domain models are in a constant state of evolution. Without the bedrock of a stable suite of tests, it can become very hard to sustain a reliable process. Approaches such as the test pyramid, testing trophy, honeycomb, and so on are acknowledged as reasonable ways to implement a sound continuous testing strategy. All of these approaches are based on the premise that a large number of cheap (computationally and cognitively) unit tests form the foundation of the strategy, with the number of tests in other categories (service, UI, manual, and so on) reducing as we move through the chain.

However, we are in this new world of fine-grained components that work by communicating with each other. Hence, there is a bigger need to verify interactions at the periphery in a robust manner. Unit tests alone that rely mostly on mocks and stubs may not suffice because the behavior of collaborators may change inadvertently. This may lead to a situation where unit tests may run successfully, but the overall functionality may be broken. This may cause teams to lose faith in the practice of unit testing as a whole and resort to using more end-to-end functional tests. However, these styles of tests can be extremely expensive (`https://www.youtube.com/watch?v=VDfX44fZoMc`) to set up and maintain, especially when we are looking to automate them. Consequently, most teams ignore the results of a majority of automated testing methods and rely almost exclusively on manual testing to verify anything but the most trivial functionality.

Any manual testing requires most if not all functionality to be ready before any meaningful testing can commence. Furthermore, it is time-consuming, error-prone, and usually not repeatable. Consequently, almost all testing can be carried out only when it is very close to the end, rendering the idea of continuous testing a pipe dream. Despite all its limitations, teams continue to rely on manual testing because it seems to provide the most psychological safety in comparison to its automated counterparts.

In an ideal world, what we need is the speed of unit tests and the confidence provided by manual testing. We will look at a few specific forms of testing that can help restore the balance.

Contract testing

The limitation of unit testing is that the assumptions made in mocks/stubs can be invalid or become stale as producers make changes to the contract. On the other hand, manual tests suffer from being slow and wasteful. Contract tests can provide a means to bridge the gap by providing a happy medium where the producer and consumer share an executable contract that both producer and consumer can rely on as functionality changes/evolves. At a high level, this works in the manner depicted here:

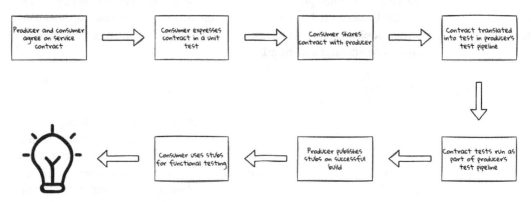

Figure 12.2 – Contract testing: high-level flow

This allows the consumers and the producers to work collaboratively, and get feedback a lot earlier in the cycle. For the consumer, they get to participate in sharing their expectations with the producer and make use of versioned, producer-approved stubs for their own testing without having to depend on the producer's real system. Likewise, producers gain a deeper understanding of how their services are consumed, setting them free to make bolder changes as long as they remain compatible.

Test-First Design

The essence of domain-driven design is all about gaining as thorough an understanding of the problem in order to solve the problem right. Test-first design enables gaining a better understanding of the problem because it mitigates the risk of becoming biased by the solution we have built. In addition, it also promotes the automated verification of these requirements, which allows them to be used as an effective aid to regression testing. We are strong proponents of this practice for this reason and encourage you to consider adopting TDD as a core practice to accentuate your effectiveness with DDD.

Mutation testing

A lot of teams author a variety of tests to ensure that they are building a high-quality solution. Test coverage is typically used as a quantitative measure to assess the quality of testing. However, test coverage is a necessary but not sufficient condition to establish test quality. Low test coverage almost definitely means there is a test quality problem, whereas high coverage does not imply better tests. In an ideal world, even a single line change in production code (caused by a change in business requirements), without changing test code, will result in a test failure. If this can be guaranteed for every single change across the code base, you may be able to safely rely on such a test suite.

Mutation testing is a practice that automatically inserts small bugs in production code (called *mutants*) and reruns an existing suite of tests to ascertain the quality of tests. If your tests failed, the mutant is killed. Whereas if your tests passed, the mutant survives. The higher the number of mutants killed, the more effective your tests are.

For example, it may apply mutations such as inverting conditionals, replacing relational operators, returning nulls from methods, and so on, and then you can check the effect this has on your existing tests. If no tests fail despite these mutations, these tests may not be as helpful as you hoped them to be. This allows us to draw more objective conclusions about the quality of our tests. Given how it works (by mutating code), it is computationally intensive and hence may take a long time to run. If you employ a test-first design and have a fast suite of unit tests, mutation testing can be a great complement that can help discover missed requirements and/or test cases earlier in the development cycle. From that perspective, we see it as an invaluable tool to augment the adoption of DDD within teams. Tools such as PITest (`https://pitest.org/`) are a great tool to perform mutation testing in your Java applications.

Chaos testing

As we have seen earlier, mutation testing can help point out chinks in the functional aspects of your application. Chaos testing plays a similar role to help identify shortcomings in meeting non-functional requirements caused by reliance on network and infrastructure. It started becoming popular through the use of large-scale distributed, cloud-based architectures pioneered by companies such as Amazon, Netflix, and so on. Netflix initially released a tool called Chaos Monkey (`https://netflix.github.io/chaosmonkey/`) that randomly terminated instances in production(!) to ensure that engineers implement services that are resilient to failure. They followed this by releasing a set of related tools, collectively called the Simian Army (which is now defunct) to test a variety of non-functional aspects such as latency, security compliance, unused resources, and so on.

While Netflix performs this style of testing in production, the rest of us will benefit immensely if we adopt these practices even in lower environments at the outset. From a strategic perspective, chaos testing can provide feedback on the amount of coupling between components and whether the boundaries of these components are appropriate. For example, if a component that you are dependent on goes down or experiences problems, does this take you down as well? If so, are there ways to mitigate this? It can also provide feedback about your monitoring and alerting apparatus. From a tactical perspective, it can provide insights into the shortcomings of the invocation style being used to communicate among components.

In this section, we have chosen to highlight contract testing, mutation testing, and chaos testing because we see them as game-changers in the application of DDD. Teams will benefit by looking at these methods as augmentations to other testing methods when coming up with a well-rounded testing strategy.

Deployment automation

The intent of applying domain-driven design is to create an ecosystem of loosely coupled components – so that each of these components can evolve independently of each other. This includes how these components are deployed to production. At a high level, we have at least three styles of deployment:

- **Single-process monolith**: Where large portions of the application are deployed as a single unit, with all components that are included in the deployment running in a single process

- **Distributed monolith**: Where the application is split into multiple components with each running in its own process and/or host, but deployed as a single unit and/or requiring non-trivial amounts of coordination and tight coupling among components and their owners

- **Independent components**: Where the application is split into multiple components with each running in its own process and/or host, deployed independently of each other and requiring minimal to no coordination among component owners

We also have a number of deployment strategies that we can employ. We list some of the more popular ones in order of increasing complexity and richness:

- **Basic**: Likely the oldest style of deployment where the newer version of the application replaces the old, usually with some amount of downtime. Rollback typically means redeploying the previously live version, again taking some amount of downtime. This is a fairly common deployment strategy for those applications where a certain amount of downtime is acceptable. This may include non-business critical applications and/or third-party packages where we do not have a say in how those applications manage their deployments. In the case of certain monoliths, this may be the only feasible option due to the overall complexity of the system as a whole. This style of deployment typically starts out being fairly simple and well understood and may suffice for non-critical applications. On the flip side, it requires the deployment and release to happen in one single tightly coupled step and may involve some amount of downtime.

- **Blue-green**: A deployment strategy that makes use of two identical environments, a "blue" and a "green" environment, with one representing the current production and another representing the newer version. The current version continues to service traffic, while testing and acceptance are carried out on the new version without exposing it to end users. User traffic is switched to the newer version once testing activities are deemed to be successfully completed. It is pertinent to note that live user traffic is directed only to one environment at any given time. This style of deployment enables deployment with (near) zero downtime and also allows decoupling of the process of deployment and release. Rollbacks are easier because it simply means redirecting traffic to the older version. On the other hand, it requires double the amount of capacity at least during the time of deployment. This may make it cost-prohibitive for monolithic applications.

- **Rolling**: A deployment strategy where a small subset of current version instances is incrementally replaced by newer version instances. Both old and new versions of the software continue to run in parallel until all instances of the old are replaced with new ones. In simple cases, rollback typically means replacing the newer version instances with older ones. This style of deployment also enables zero-downtime deployment, while also allowing side-by-side testing of old and new versions with real users. Rolling deployments can make rollbacks relatively easy by aborting the introduction of instances of the new version and re-introducing the old version and hence can reduce the *blast radius* of a bad release. Unlike the case with blue-green deployments, here deployment and release cannot be decoupled. Deployment means that the system is released (at least for a subset of users).

- **Canary**: A variation of the rolling deployment where traffic is routed to newer instances in a controlled and phased manner, typically an increasing proportion of request volume (for example, 2% → 25% → 75% → 100% of users). This deployment style enables more fine-grained control of the extent of the release as compared to rolling deployment.

- **A/B deployment**: A variation of canary deployment where multiple versions (with one or more variations) of new functionality may run simultaneously as "experiments" along with the current version. Further, these variations may be targeted to specific sets of users. It allows for testing more than two combinations at the same time with real users.

When working with monolithic applications, teams are usually forced to restrict themselves to either basic or at the most blue-green deployments because the cost and complexity involved in adopting more sophisticated deployment strategies are a lot higher. On the other hand, distributed monoliths make this even more complicated because it now requires coordination among physically disparate components and teams. As long as we are able to maintain a balance between component granularity and coupling, we should be able to support a variety of advanced deployment strategies.

In today's modern ecosystem where there is a tremendous amount of competition to deliver new features and innovate faster, there is a need to support more complex forms of deployment with the least amount of risk and disruption to the business. If supporting flexible deployment strategies proves to be too hard, there is very likely a need to re-examine your context boundaries.

Refactoring

Over a period of time, there will be a need to realign context boundaries, domain events, APIs, and so on. There tends to be a stigma associated with things not working perfectly the first time and justifying the need for refactoring at the inter-component scale. However, this may be required for multiple reasons outside our control, ranging from competitor ecosystem changes, evolving/misunderstood requirements, inability to meet non-functional requirements, organizational and team responsibility changes, and so on. Hence, refactoring is a core discipline that software teams will need to embrace as a first-class practice.

> **Note**
>
> We are covering only the strategic (inter-component) aspects of refactoring in this chapter. There are several great works on the tactical (intra-component) aspects of refactoring, such as Martin Fowler's *Refactoring* (`https://refactoring.com/`) book and Michael Feathers' *Working Effectively with Legacy Code*, among others.

From a strategic perspective, this may mean having to break an existing monolith into finer-grained bounded contexts or merge fine-grained bounded contexts into more coarse-grained ones. Let's look at each of these in turn.

Breaking an existing monolith

In previous chapters (10 and 11), we have looked at how it is possible to break an existing monolith into finer-grained components. However, it is arguable that the monolith was relatively well-structured to start with. Lots of teams may not be as fortunate. In such a case, here are some prerequisites that may need to be fulfilled:

- **Perform tactical refactorings**: This will allow you to gain a better understanding of the existing system. To do this, start with a set of fitness functions (`https://en.wikipedia.org/wiki/Fitness_function`) and a set of black-box functional tests, perform a refactor, and then replace the functional tests with faster-running unit tests. Finally, use the fitness functions to evaluate the success of the effort. Repeat this process until there is a level of comfort to attempt more complex refactorings.

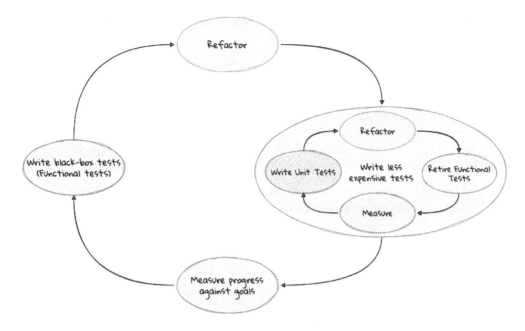

Figure 12.3 – Continuous improvement loop

- **Introduce domain events**: Identify software seams (http://wiki.c2.com/?SoftwareSeam) and publish domain events along those seams. Use the domain events to start decoupling the producers and the consumers.

- **Pick low-hanging components**: If possible, pick areas with low afferent coupling and low to medium complexity at the outset. This will allow you to get a firmer grasp of applying these techniques before attempting more complex ones. Please refer to *Chapter 10, Beginning the Decomposition Journey* and *Chapter 11, Decomposing into Finer-Grained Components* for details on how to proceed.

Merging into coarse-grained bounded contexts

Merging two distinct bounded contexts can be relatively less complex than breaking down an existing one. However, there are a few nuances that are worth paying attention to, in the following order:

- **Unification of the ubiquitous language**: In *Chapter 9, Integrating with External Systems*, we examined a variety of ways in which bounded contexts can integrate with each other. If the relationship between these bounded contexts is symmetric, there may be less work to do. This is because, in a symmetric relationship, there likely exists a lot of synergies in the first place. However, if the relationship is asymmetric, for example, through an open-host service on the producer side and an anti-corruption layer on the consuming side, it means that there are possibly two varying ubiquitous languages and likely distinct domain models at play. Careful thought will need to be applied to arrive at a ubiquitous language that is applicable across the newly merged bounded context.

- **Adjust internal domain models**: Adoption of a common ubiquitous language primarily means making use of a common domain model across the newly merged bounded context. This means that the aggregates, entities, and value objects will need to be unified, which may then require changes at the persistence layer as well. If there are domain events that are published and consumed exclusively between these components, those domain events may be candidates to be retired. At this stage, it may not be prudent to make any changes to any public interfaces – specifically those exposed using an open-host service (for example, public HTTP APIs and other domain events).

- **Adjust the public API design**: As a final step, it will be prudent to refactor redundant and/or inefficient public interfaces to conclude the exercise and derive the intended benefits.

It is pertinent to note that this style of continuous improvement can be extremely challenging to adopt without the solid bedrock of a sound set of engineering practices, specifically the testing and deployment automation practices that we discussed in this section.

Invocation style

When integrating two bounded contexts that are running in distinct processes, there are two ways to consummate interactions: synchronous and asynchronous.

Synchronous invocation

The client blocks until the server provides a response. Optionally, implementations can choose to wait for an amount of time for the invoked operation to complete before timing out. An example of such an interaction is a blocking HTTP call made to start a new LC application like so:

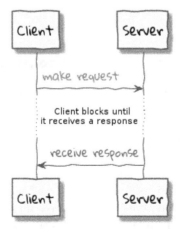

Figure 12.4 – Synchronous invocation

When the call returns successfully, the client is sure that their request to create a new LC application has worked. If the server is slow to respond, it can result in a performance bottleneck, especially in high-scale scenarios. To cope with this, the client and the server may agree on a response time SLO for that interaction. The client can choose to wait for a response from the server for the agreed amount of time after which the client times out the request and considers it a failure. Given that the client blocks on a server response, it is not able to do anything else while it waits, even though it may have the resources to do other things. To deal with this, the client can employ an asynchronous invocation.

Asynchronous invocation

In an asynchronous style of invocation, the client interacts with the server in a manner that frees it to perform other activities. There are a few ways to do this:

- **Fire and forget**: The client initiates a request with the server, but does not wait for a response from the server and also does not care about the outcome. Such a style of interaction may suffice for *low-priority* activities such as logging to a remote server, push notifications, and so on.

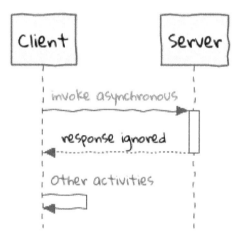

Figure 12.5 – Fire and forget

- **Deferred response**: In some (many?) cases, the client may need to know the outcome of the request they had previously made. If the server supports it, the client can submit a request, just wait for a confirmation that the request was received along with an identifier of the resource to be tracked, and then poll the server to track the status of its original request as shown here:

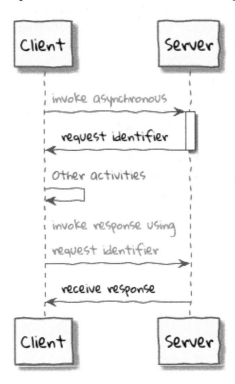

Figure 12.6 – Deferred response using poll

- **Request with callback**: When the client polls for the response, the server may not be finished with processing the original request. This means that the client may need to poll the server more than once to understand the status of the request, which can be wasteful. An alternative is for the server to push a response back to the client when it has finished processing by invoking a callback that the client provided when making the request.

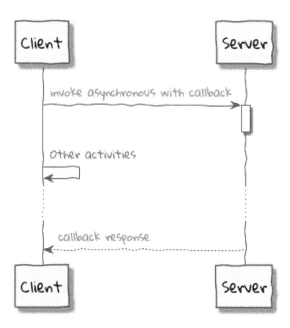

Figure 12.7 – Deferred response using callback

Given that these interactions happen over a network that can be unreliable, clients and servers need to employ a variety of techniques to achieve some semblance of reliability. For example, clients may need to implement support for timeouts, retries, compensating transactions, client-side load balancing, and so on. Similarly, the server may need to protect itself from errant clients by making use of techniques such as rate limiters, circuit breakers, bulkheads, fallbacks, health endpoints, and so on.

> **Note**
>
> Elaborating on the specific techniques mentioned here is out of scope for this book. Books such as *Release It* and *Mastering Non-Functional Requirements* cover these patterns in a lot more depth.

In a lot of cases, there is usually a need to employ a combination of several of the preceding techniques to provide a resilient solution. Just as we discussed in the logging section, mixing these concerns with core business logic can obscure the original intent of the problem. In order to avoid this, it is advisable to apply these patterns in a manner that is peripheral to core business logic. It may also be prudent to consider the use of libraries such as `Resilience4j` (`https://resilience4j.readme.io/`) or `Sentinel` (`https://github.com/alibaba/Sentinel`).

Logging

Application logging is one of the most fundamental aids when it comes to diagnosing issues in running code. In a lot of code bases, logging tends to be an after-thought where developers add log statements only after they encounter problems. This results in log statements being strewn almost randomly throughout the code base. Here is a simple example of code within a command handler to log its execution time among other things:

```
@Log4J2
class LCApplication {
    //...
    @CommandHandler
    public LCApplication(StartNewLCApplicationCommand command) {
        log.debug("Starting execution of command with applicant {}",
                    command.getApplicantId());

        long timeTaken = -System.nanoTime();

        log.debug("Appending event for command {}", command.getApplicantId());

        AggregateLifecycle.apply(
                new LCApplicationStartedEvent(command.getId(),
                command.getApplicantId(),
                command.getClientReference(), LCState.DRAFT));

        log.debug("Appended event for command {}",
                    command.getApplicantId());

        timeTaken += System.nanoTime();
        log.debug("Ending StartNewLCApplication in {} nanoseconds.",
timeTaken);
    }
}
```

There is no doubt that this logging code can be invaluable when troubleshooting issues. However, when we look at the preceding code, the logging code seems to dominate the entire method obscuring the domain logic. This might feel innocuous, but when this is done in multiple places, it can get quite repetitive, cumbersome, and error-prone – compromising readability. In fact, we have seen cases where seemingly innocent log statements have introduced performance issues (for example, within a loop with an expensive argument evaluation) or even bugs (for example, the dreaded NullPointerException when trying to evaluate arguments). In our opinion, it is very important to treat logging as a first-class citizen and afford it the same rigor as core domain logic. This means that it needs to obey all the good practices that we associate with well-factored production code.

Segregating logging code

Ideally, we will be able to maintain a balance between readability and debuggability. This can be achieved if we can segregate these two concerns. One way to segregate this cross-cutting logic is to use aspect-oriented programming (read more about AOP at `https://www.eclipse.org/aspectj/` and `https://docs.spring.io/spring-framework/docs/current/reference/html/core.html#aop`) as shown here:

```
1  @Aspect
2  @Component
3  public class TimingAspect {
4
5      @Around("@annotation(org.axonframework.commandhandling.CommandHandler)")  ❶
6      public Object log(ProceedingJoinPoint point) {
7
8          logEntry(point);
9
10         long time = -System.nanoTime();
11         boolean normalExit = false;
12         try {
13             final Object out = point.proceed();
14             normalExit = true;
15             return out;
16         } finally {
17             logExit(point, normalExit, time);
18         }
19     }
20 }
```

> **Note**
>
> A pointcut defines an `around` aspect for all methods annotated with the `@CommandHandler` annotation. In this example, we are using compile-time weaving as opposed to runtime weaving available through the Spring Framework, to inject execution time logic using AspectJ. You can find more details on the pros and cons of using specific weaving techniques in this article (`https://www.baeldung.com/spring-aop-vs-aspectj`).

In the style shown here, we have separated logging code from application code through the use of aspect-oriented programming. In the example, the logging code applies to all methods annotated with the `@CommandHandler` annotation. This has the advantage that all such methods will now produce consistent entry/exit logging statements. On the flip side, if there is a need for additional logging for a specific command handler, it will still have to be done within the body of that method. If you see yourself requiring lots of ad hoc logging statements in addition to simple entry/exit logs, it might be a smell and a sign that your methods may need to be refactored.

Dealing with sensitive data

In general, when adding logging code, it helps to include as much context as possible. This can be challenging in certain domains such as healthcare or finance where there may be legal/regulatory requirements to restrict access to sensitive information. For example, during the LC application process, we may need to perform a credit check for the applicant using their government-issued identifier such as the **social security number** (**SSN**) in the USA. In such cases, it is common to mask a significant portion of this information in the logs to maintain a balance between privacy and debuggability. Implementing such masking logic is a domain concern. This accentuates the need to use value types (as opposed to primitives) to control the appropriate behavior. For example, overriding the `toString` method of an `SSN` value type can ensure that the sanctity of the business need is met uniformly within the bounded context.

While masking may suffice in a majority of use cases, it suffers from the limitation of not being able to access the original information even by authorized users. If this is a requirement, it may be necessary to make use of **tokenization** (the process of replacing a sensitive piece of information with a non-sensitive placeholder value called a **token**) solution. This can allow logging tokenized values in an unrestricted manner within the bounded context and in general, can be a lot more secure. But this can mean having to deal with the additional complexity of another bounded context to provide tokenized values and authorization controls when the real value needs to be accessed.

Log format

Thus far, we have focused on just the log message. However, logging is more than just that. It is typical to include additional information such as the time of occurrence, log level, and so on to aid in rapid troubleshooting. For example, Spring Boot uses the following log format by default:

```
2022-06-05 10:57:51.253  INFO 45469 --- [ost-startStop-1] c.p.lc.
app.domain.LCApplication              : Root WebApplicationContext:
Ending StartNewLCApplication in 1200495 ns.
```

While this is an excellent default, it still is primarily unstructured text with certain information being lost in order to improve readability (for example, the logger name is abbreviated). While logs are primarily meant to be consumed by humans, a large volume of logs can get in the way of being able to locate the relevant logs. So it is important to produce logs that are also machine-friendly so that they can be easily indexed, searched, filtered, and so on. In other words, using a *structured logging* format like the one shown next can go a long way toward meeting the goals of both machine and human readability:

```
 1   {
 2       "@timestamp": "2022-06-23T18:03:49.135+01:00",
 3       "@version": "1",
 4       "acceptHeader": "application/json",
 5       "level": "INFO",
 6       "level_value": 20000,
 7       "logger_name": "com.premonition.lc.application.domain.LCApplication",
 8       "message": "Ending StartNewLCApplication in 1200495 ns",
 9       "thread_name": "worker-1"
10   }
```

Making use of a structured log format elevates their use from being just a debugging tool to becoming yet another rich and cheap source to derive actionable business insights.

> **Note**
>
> While it may be tempting to pick a custom log format, we strongly recommend picking formats that are compatible with popular ones such as Apache's **Common Log Format** (**CLF**) (`https://httpd.apache.org/docs/current/logs.html#common`) or Logstash's default format (`https://github.com/logfellow/logstash-logback-encoder#standard-fields`).

Log aggregation

The fact that our applications are decomposed into multiple components with each usually running multiple instances means that this can produce a lot of logs that are disconnected from each other. To be able to work with these logs meaningfully, we need to aggregate and sequence them chronologically. It may be worth considering the use of a formal log aggregation solution for this purpose. Using a structured logging solution as previously discussed can go a long way when working with logs from multiple systems.

> **Note**
>
> For more information on logging best practices, please refer to this logging cheatsheet from OWASP (`https://github.com/OWASP/CheatSheetSeries/blob/master/cheatsheets/Logging_Cheat_Sheet.md`) and also this article on the art of logging (`https://www.codeproject.com/Articles/42354/The-Art-of-Logging`).

Aggregating logs in one place allows us to view diagnostic information from multiple applications. However, we still need to correlate this information when in the midst of a flow. Distributed tracing solutions can help with this. Let's look at this next.

Tracing

Imagine a situation where an applicant submitted an LC application through the UI. When all goes well, within a few milliseconds, the applicant should get a notification of successful submission as shown here:

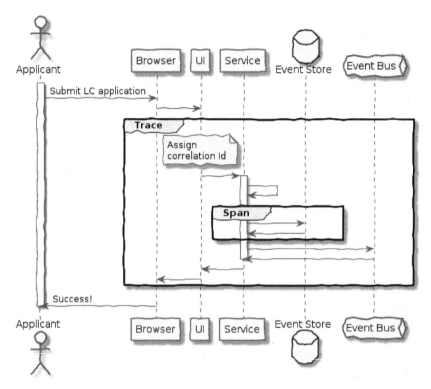

Figure 12.8 – Submit LC application flow

Even in this simple example, there are several components involved, each of which produces logs of its own. When an engineer is looking to diagnose an issue, there is a need to correlate log entries from multiple components. In order to accomplish this, there is a need to introduce a correlation identifier as close to the start of the interaction and propagate it across component boundaries. Furthermore, log entries in each component need to carry this correlation identifier as they produce logs. Doing this will allow us to view log entries spanning process boundaries using the correlation identifier as a unifying thread. In technical terms, the entire flow is called a *trace*, and each segment within the flow is called a *span*. This process of instrumenting log entries with such information is termed *distributed tracing*.

As is evident here, user flows may – and usually do – span more than one bounded context. For this to work effectively, bounded contexts need to agree on propagating trace and span identifiers uniformly. Tools such as Spring Cloud Sleuth and OpenTracing can help simplify implementation for teams using disparate technology stacks.

Fundamentally, distributed tracing visualizations can aid in diagnosing performance bottlenecks and chattiness between components. But what may not be obvious is the insights it can provide in gaining a richer understanding of how components interact in an end-to-end user journey. In a lot of ways, this can be thought of as a near real-time context map visualization of your system, and how components are coupled with each other. From a DDD perspective, this can provide greater insights into re-evaluating bounded context boundaries if necessary. For this reason, we strongly recommend making it easy to set up and configure distributed tracing apparatus painlessly right from the outset.

Versioning

When we are working with a monolithic application, we have large portions bundled as a single cohesive unit. This means that other than third-party dependencies, we don't have to worry about explicitly versioning our own components. However, when we start breaking components into their individual deployable units, there is a need to pay careful attention to how the components, APIs, and data elements of our solution are versioned. Let's look at each in turn.

Components

When we create components, there are two broad categories – those that are **deployed** on their own and those that are **embedded** within another component. In the case of deployable components, there is a need to use an explicit version to identify specific instances of the component, even if only for deployment purposes. In the case of the embedded component, again there is a need to use an explicit version because other components need to understand what instance they depend upon. In other words, *all components* need to have a version to uniquely identify themselves.

It follows that we then need to choose a sound versioning strategy for our components. We recommend the use of semantic versioning (`https://semver.org/`), which uses a version identifier that uses three numeric components that match the **MAJOR.MINOR.PATCH** scheme:

- **MAJOR**: Increment when you make backward-incompatible changes.
- **MINOR**: Increment when you add functionality in a backward-compatible manner.
- **PATCH**: Increment when you make backward-compatible bug fixes.

In addition, we can make use of optional extensions to indicate pre-release and build metadata. For example, the version identifier for our component might read 3.4.1-RC1 to reflect that this is a release candidate for version 3.4.1 of our component. Using a standard versioning scheme enables the use of build tools such as Maven and Gradle to declare fine-grained upgrade rules and constraints for direct and transitive dependencies. A good practice here is to declare dependencies without versions and make use of dependency management (`https://maven.apache.org/guides/introduction/introduction-to-dependency-mechanism.html#Dependency_Management`) or dependency constraints (`https://docs.gradle.org/current/userguide/dependency_constraints.html#sec:adding-constraints-transitive-deps`) to centralize version management of dependent components.

APIs

As producers, we expose APIs in a number of ways. In this case, we are specifically referring to APIs made available over remote interfaces such as HTTP, events, and so on. When it comes to APIs, first and foremost, it is important to keep consuming applications functionally. One effective way of making this possible is by thinking from the consumer's standpoint and embracing consumer-driven contracts (`https://martinfowler.com/articles/consumerDrivenContracts.html`).

From a consumer's perspective, the robustness principle (Postel's law) applies: *be conservative in what you send, be liberal in what you accept*. In other words, when sending requests to providers, strictly obey the constraints laid down by the producer. For example, don't send unexpected data in the request. Whereas, when receiving responses, be tolerant towards what you get from the producer. For example, ignore unknown attributes in the response as long as all the attributes you expect are present. This will allow producers to evolve without breaking existing consumers.

Our recommendation is to keep APIs versionless for as long as possible by continuing to maintain backward compatibility. Despite all our efforts, there may come a need to make breaking changes to our APIs. Breaking changes include the following:

- Removing/renaming one or more attributes
- Changing the type of one or more existing attributes
- Changing the format of the request/response

In such cases, make use of a version identifier to indicate major version changes (for example, v2 to v3). Common options include specifying the version in the URI, in a header, or in the payload. But as we have mentioned earlier, API versioning needs to be used sparingly. If you find yourself in a situation where you are required to introduce backward-incompatible changes frequently, it might be an indicator of requirements being misunderstood and whether DDD principles are truly being applied.

Data

In a world of well-defined bounded contexts, we should no longer be in a situation where we need to expose data directly to our consumers. However, there may be situations where we may need to integrate by directly exposing data to our consumers. For example, we may have to expose a reporting database for analytical purposes. All the good practices that we outlined for APIs apply to data as well.

In addition, from a producer's perspective, there will be a need to evolve the data schema to accommodate changing business requirements. When working with relational databases, using good schema migration tools such as Liquibase (`https://liquibase.org/`) or Flyway (`https://flywaydb.org/`) can go a long way. NoSQL databases also have similar tools such as MongoBee (`https://github.com/mongobee/mongobee`) and Cassandra-Migration (`https://cassandra.tools/cassandra-migration`).

In this context, it is pertinent to think about data as a product and apply product thinking to domain-aligned data. For more information, please refer to this article on how to move from a monolithic data lake to a distributed data mesh (`https://martinfowler.com/articles/data-monolith-to-mesh.html#DomainDataAsAProduct`).

It is not uncommon to find ourselves in situations where there may be a need to support more than one active version of a given component, API, or data. This can add significant levels of complexity to the solution. To keep complexity in check, it is important to make provisions for deprecating and eventually ending support for older versions.

Summary

In this chapter, we looked at aspects purely beyond functional requirements – each of which can have a profound impact on our ability to apply domain-driven design effectively. Specifically, we looked at how each of these is interrelated and they have to be looked at holistically to achieve and sustain high levels of success.

Closing thoughts

Domain-driven design, although conceived in the early 2000s, was way ahead of its time. We are in the age of solving the most complex problems yet. Given the advancements in technology, there is an expectation to build these solutions a lot faster. While the overall cognitive complexity of the solution is directly proportional to the complexity of the problem, there is a need to effectively manage this complexity. DDD and its principles enable us to achieve this by breaking down complex problems into smaller, manageable parts. In this book, we have made an attempt to distill our experiences and provide a set of concrete techniques to apply DDD in your respective contexts.

Index

Other Books You May Enjoy

If you enjoyed this book, you may be interested in these other books by Packt:

Hands-On Software Architecture with Java

Giuseppe Bonocore

ISBN: 978-1-80020-730-1

- Understand the importance of requirements engineering, including functional versus non-functional requirements

- Explore design techniques such as domain-driven design, test-driven development (TDD), and behavior-driven development

- Discover the mantras of selecting the right architectural patterns for modern applications

- Explore different integration patterns

- Enhance existing applications with essential cloud-native patterns and recommended practices

- Address cross-cutting considerations in enterprise applications regardless of architectural choices and application type

The Complete Coding Interview Guide in Java

Anghel Leonard

ISBN: 978-1-83921-206-2

- Solve the most popular Java coding problems efficiently

- Tackle challenging algorithms that will help you develop robust and fast logic

- Practice answering commonly asked non-technical interview questions that can make the difference between a pass and a fail

- Get an overall picture of prospective employers' expectations from a Java developer

- Solve various concurrent programming, functional programming, and unit testing problems

Packt is searching for authors like you

If you're interested in becoming an author for Packt, please visit authors.packtpub.com and apply today. We have worked with thousands of developers and tech professionals, just like you, to help them share their insight with the global tech community. You can make a general application, apply for a specific hot topic that we are recruiting an author for, or submit your own idea.

Share Your Thoughts

Now you've finished *Domain-Driven Design with Java - A Practitioner's Guide*, we'd love to hear your thoughts! Scan the QR code below to go straight to the Amazon review page for this book and share your feedback or leave a review on the site that you purchased it from.

https://packt.link/r/1800560737

Your review is important to us and the tech community and will help us make sure we're delivering excellent quality content.

www.ingramcontent.com/pod-product-compliance
Lightning Source LLC
Chambersburg PA
CBHW062111050326
40690CB00016B/3289